D0743098

HOW TO MAKE

PRINTED CIRCUIT BOARDS

With 17 Projects

To Teri Lynn
The joy of my life

No. 2898
$23.95

HOW TO MAKE

PRINTED
CIRCUIT BOARDS

With 17 Projects

CALVIN GRAF

TAB BOOKS Inc.
Blue Ridge Summit, PA

FIRST EDITION
FIRST PRINTING

Copyright © 1988 by Calvin R. Graf
Printed in the United States of America

Library of Congress Cataloging in Publication Data

Graf, Calvin R.
    How to make printed circuit boards, with 17 projects / by Calvin
R. Graf.
        p.   cm.
    Includes index.
    ISBN 0-8306-9098-0   ISBN 0-8306-2898-3 (pbk.)
    1. Printed circuits. 2. Printed circuits—Design and construction. I. Title.
TK7868.P7G76 1988
621.381'74—dc19   88-20132
                                                              CIP

TAB BOOKS Inc. offers software for sale. For information and a catalog, please
contact TAB Software Department, Blue Ridge Summit, PA 17294-0850.

Questions regarding the content of this book
should be addressed to:

    Reader Inquiry Branch
    TAB BOOKS Inc.
    Blue Ridge Summit, PA 17294-0214

Front cover photograph courtesy of Tektronix, Inc.

CONTENTS

Preface   vii

Acknowledgments   ix

Introduction   x

1   General Principles of Electronic Construction   1

Getting Started—The Home Worker—The Shop Worker—Design or Build?—
Electronic Project Kits—Neatness and Workmanship—Workshop Hints—Care in
Using Tools and Equipment—Safety on the Job

2   Electron Theory   13

The Atom . . . What Is It?—Ring Around the Atom—Metal Conductors—The
Insulator—The Semiconductor—The Doped Crystal

3   Electronic Components   23

Types of Electronic Components—Electrical Properties—Electronic Components—Semiconductor Devices—Electrical Units and Symbols

4   Electronic Diagrams   41

Block Diagrams—Pictorial Diagrams—Layout Diagrams—Wiring Diagrams—
Schematic Diagrams

5 Electronic Tools and Soldering   **56**

Hardware — Tools in General — Soldering Tools — Soldering — Desoldering — Techniques

6 Types of Wiring and Circuit Boards   **86**

Wire Designations — Wire Sizes — Point-to-Point Wiring — Breadboarding — Perforated Boards — Printed Wiring — Component Mounting

7 Circuit Design and Board Layout   **106**

Printed Circuit Board Layout — Discrete Component Layout — Transferring Schematic to Board — A Review Circuit — The Printed Circuit Board Process — Commercial Board Kits — Direct-Etch PCB for Simple Circuits — TEC-200 Image Film — DASH PCB Layout — Drilling the Board — Mounting and Soldering Components — Board Clean-up and Testing

8 Commercial Assembly Kits   **127**

Electronics Kits International — Radiokit — Radio Shack Kits — Dick Smith Electronics — Jameco Electronics — Graymark Komponent Kits — Heathkit/Zenith Kits

9 Build These Practical Projects   **152**

Fields of Application — Brief Project Descriptions

Projects
Audio Continuity and Voltage Tester 155
Continuity/Polarity Tester and LED Power Supply 157
Intrusion Alarm Trip Circuit 163
Noise-Activated Switch and Piezo Sounder 166
Annunciator Sounder 168
Whistler Lightning Receiver 169
Light Sensor Alarm 173
Multiprobe Piezo Sounder 176
Resistance-Sensitive Audio Continuity Tester 179
Headlight Reminder 181
Lamp Dimmer and Motor Control 184
Ultrasonic Transmitter and Receiver 185
Nightlight Turn-On 189
Low-Level Audio Amplifier 191
Battery Eliminator 193
Telephone On-Hold Project 195
Visual Telephone Ringer 197

Index   **203**

PREFACE

Space travel of man and the walk on the moon would not have been possible without advancements in primarily one area—technology. This involves the technology of rocket propulsion, radio communications, and space navigation. Radio communications and space navigation depend entirely on electronics, and modern electronics depends entirely on two great strides in technology: the invention of the transistor, and the development by industry of the printed circuit board (also known as the printed wiring board or PC card).

Several familiar phrases have been uttered, thanks to the small, lightweight, printed circuit board that permitted man to begin his quest to reach for the stars . . . "Liftoff. We have liftoff!" . . . "A-OK" . . . and "Tranquility Base here. The Eagle has landed!"

Before space travel, there was the transistor radio that you could slip into your shirt pocket because it was so small. However, this would not have been possible were it not for the PCB that enabled the electronics manufacturing industry to take full advantage of the small size of the transistor. And when the integrated circuit chip (IC chip) came along, it too fell right into place on the PCB. And it looks like the PCB will continue to be used in the future until the copper foil used on the board, which carries the electrons as current, is replaced by fiberoptic strands, or cables, which carry photons of light instead of electrons.

Because the magic day of the use of fiberoptics for signaling is a bit off in the future, the use of the printed circuit board dominates all electronics circuit assembly. That is what this book is about—the printed circuit board. You will understand how to lay it out, how to etch it, how to "stuff" it with component parts, and how to build your own projects. For some readers, it could be that "one small step for a man" as transmitted to the world from the moon by Neil Armstrong. For some, it could be "a giant leap for mankind!" Good luck in your venture!

ACKNOWLEDGMENTS

There are a number of sources who contributed to the success of this book. I gratefully acknowledge the fine cooperation of the listed firms. Their kind response enables many experimenters, technicians, electronics enthusiasts, science-fair entrants, newcomers to the amazing and interesting field and hobby of electronics, and skilled engineers who might be trying their first hand at "rolling their own" circuit projects to be more fully aware of the great satisfaction that comes from designing and assembling their own printed circuit boards and projects.

AP Products, Inc.
Electronics Kits International, Inc.
Future Net Corporation
GC Electronics
Graymark
Heath Company
Jameco Electronics
JimPak Electronics Components

J.J. Blair Associates, Inc.
Meadow Lake Corporation
OK Machine and Tool Corp.
PACE, Incorporated
Plato Products, Inc.
RADIOKIT
Radio Shack
Dick Smith Electronics

INTRODUCTION

COMPONENTS

Many devices and appliances in use every day in our homes, offices, businesses, autos, schools, airplanes, satellites, space and on the moon are possible only because of one thing that holds all electronic devices together—the printed circuit board, or PCB.

The chapters that follow cover a myriad of subjects that describe circuit board construction, electronic projects that are available commercially, and a number of projects that you can build and assemble yourself.

Chapter 1 discusses the general principles of electronics construction, how to get started, the use of tools, and shop safety. Chapter 2 covers electronics theory, starting with the simple and smallest electrical particle—the electron. Conductors and insulators are covered in an easy-to-understand manner before the explanation of semiconductors.

Chapter 3 leads into the various types of electronic components that help shape the electronics world—the resistor, the capacitor, and the inductor. All the marvels of electronics are shaped and controlled by these simple devices. The component color code, units, symbols, and prefixes are covered in detail. Also explained in this chapter is the aging electron (vacuum) tube, semiconductor devices such as the transistor, and finally, the various types of integrated circuit (IC) chips.

Chapter 4 is an introduction to how the components of Chapter 3 are connected to form an electronic circuit or schematic diagram. Also covered are schematic symbols, block diagrams, schematic diagrams, layout diagrams, wiring diagrams, pictorial diagrams, and reference designations to let you know where you are on a diagram. Tools and electronic hardware are covered in Chapter 4.

An important consideration in electronics assembly is soldering, and this is covered in Chapter 5. The various types of solders, fluxes, soldering irons, desoldering items, and the soldering techniques used on PCBs are covered. Chapter 6 discusses the various types of circuit boards and wiring, including point-to-point and printed wiring.

Detailed circuit design and board layout are covered in Chapter 7. This includes the PCB process of inking and taping techniques, etching the board, drilling the board, mounting and soldering components to the board, and testing and troubleshooting the final board.

Commercially available assembly kits are covered in detail in Chapter 8. Several hundred kits that are available from local electronics stores by mail (or a 1-800 telephone number) are discussed in Chapter 8. Virtually all of these easy-to-assemble kits are inexpensive and can be put together in a short period of time.

Chapter 9 finishes up with a number of practical projects that you can readily build at home. These projects are described in detail as to how the circuits operate. Circuit diagrams and a list of parts required for the project assembly are provided. Specific details as to the parts layout and method of construction are left to the reader to put into practice what he or she has learned.

Good luck in your ventures!

CHAPTER 1

General Principles of Electronic Construction

Good work habits and practices produce good projects. In this chapter, we will look at some of the practices and techniques that make for a good electronic product, one that you will be proud to show to your friends and associates.

GETTING STARTED

Life can bring many kinds of rewards, especially when you accomplish some type of job or task that takes a special skill. An especially rewarding feeling is to complete an electronics project like the ones in these chapters, when you apply power to it, it survives the smoke test and does what it is supposed to do! That is the reward—to be satisfied with your work, with what you have been able to do, and that you can do it again.

The pride of a job well done can be satisfied in several places. This book contains electronic projects that make use of circuit boards that can be made and used at home or at your place of work as part of your daily employment. You might be a beginner in electronics or a thoroughbred in the electronic board fabrication and assembly area, but there is information in this book you will be able to use over and over.

THE HOME WORKER

When working at home, the standards are as good as your hand-iwork and talents permit. The job requirement here is to produce some electronic device, simple or complex, that performs a specific task. The electronic technician, the hobbyist, the engineer, the radio amateur, the computer enthusiast, the school science-fair entrant, the teacher or instructor, the Boy Scout merit-badge seeker, the electronic experimenter, the beginner, and the college professor all have different talent levels when it comes to electronic projects circuit design, fabrication, construction, and project use. There is something for everyone in this book. Skilled or unskilled, you'll learn as you go along. For some, it might be totally new, and for others, reinforcement.

It was the great explorer Christopher Columbus who, prior to his departure to look for a new way by sea to the East Indies, paraphrased to Queen Isabella of Spain (who was picking up the research and development tab for his trip), ''Nothing ventured, nothing gained.'' Simply put, some first-time builders at home might turn out a terrific product, but others might want to try again. But that is where the joy comes in. By continually working at your projects your skills will be refined.

THE SHOP WORKER

The electronics shop worker probably has a different outlook on his or her ability to do electronic circuit construction or fabrication. In a shop, plant, or department, the technician might be required to produce so many electronic component boards per unit time or per day. There is usually little chance for hit or miss. Often, the work is performed on an assembly line as might be done in the myriad of electronic assembly plants in this country and aboard. Usually, each worker is responsible for only a small portion of the work to be done on an electronic component board. Some workers *stuff* (mount) components on the board, some do wiring, some do soldering, and some test the boards. Seldom does one worker do all these jobs, except in a small plant or where the electronic component count on the board is low. The opportunity for advancement in any organization is presented to those workers with the most skills associated with all the work, or tasks, to be done. Therefore, it is most worthwhile for you to know and be adept in all areas of the fabrication and assembly of the electronic component boards or systems.

DESIGN OR BUILD?

The beginning electronic hobbyist or experimenter must know certain basics before designing or assembling electronic components to form a working project. I explain all of these basics in the first few chapters. Then, a number of interesting projects are described so you can choose any of them and reduce them to practice by building a model of some electronic project that works!

There are many talented professional electronics and electrical folks in the country that have never actually built an electronic project. They can look at a complicated electronic schematic, describe what each component does, and tell you what happens if a component opens or shorts out, but they have never built and experimented with an electronic project. When you get through with these chapters, you will be able to start out on the road to an extremely rewarding hobby or career.

In order to get a job as a technician, you obviously must have some knowledge about the theory of radio or electronics. You must also be able to perform many of the tasks of the technician—that of looking at a circuit and building the necessary finished product. In this case, the circuit has already been designed and is presumed to work "as advertised." The technician simply puts the circuit together, checks it out, and it (usually) works. So, in most cases, you will "build" the circuit rather than design it. However, as you become more skilled with time, you can modify a known circuit to make it do other things that *you* might want it to do. In this case, you can really say that you have modified and designed your own electronic circuit. That accomplishment will truly make you feel proud!

Take, for example, the success story of Hewlett and Packard. The Hewlett-Packard Company of Palo Alto, CA is the designer and manufacturer of a multi-billion dollar electronics industry of computers, printers, medical devices, space exploration, and the like. These fractional billionaires from the "What if?" company started their effort with $500 in a humble garage in Palo Alto. Their first product was an audio oscillator, built in the late 1930s. Walt Disney was just getting started in his color movie productions, and he bought one of the first copies of the audio oscillator to generate musical tones and sounds for his technicolor movie *Fantasia*. The movie was a tremendous hit throughout the world and is still a classic 40 years later!

ELECTRONIC PROJECT KITS

Electronic kit building is a quick and easy way to get started in electronics, or if you are a professional, it's a way to build your very own first circuit board that works. You might build the kits to help you in your hobby or to advance you in your technical field. You can buy an electronic kit to build most any device you like. The high-fidelity buff, the amateur radio operator, the computer enthusiast that wants to expand the capability of a personal computer, the radio-controlled model airplane flyer, the home safety-device experimenter, the advanced merit-badge seeker, and the science-fair entrant all have something waiting for them in kit building.

Additionally, a current trend is in building robotic devices that you can easily assemble yourself. This little-tapped field is just emerging due to the great interest in the space program as well as that of the expanding personal computer field.

Chapter 8 discusses a number of commercially available kits that can be purchased for a very nominal fee. These kits are available at your local radio parts stores or through the mail. The reward of electronic kit building is fantastic. It is both a hobby and a career field.

NEATNESS AND WORKMANSHIP

We all know how important neatness is in the things we do in our daily lives. The same should apply to your work habits in electronics. Accordingly, whether you work in industry as a career or build electronic equipment or kits as a hobby, it is very important to keep your tools and work area orderly.

Just as the properly trained auto mechanic places a work cloth or cover on the fender or work area of an automobile before beginning work, the electronics technician should also keep a neat work area. A dropped screw in an auto carburetor can cause havoc, but a dropped screw in a rotating distributor (if not retrieved) is disastrous. Accordingly, in electronic circuit assembly, a loose bit of scrap wire, some solder that has splattered, or metal filings in the wrong places can cause the equipment to malfunction. It could even cause your equipment to go up in smoke!

You will be more organized and can build with pride if you keep a clean work area. To do otherwise can dampen your work morale, invite error and frustration, and lessen your attention to detail. Therefore, your workmanship and neatness are important to the successful completion of your project. A simple rule is to keep your workbench neat and clean as you proceed with your project.

WORKSHOP HINTS

Here are very simple hints that will help you to produce a neat, working project, one you can be proud of when you show it to relatives, friends, and schoolmates. These are common-sense hints—you will think of others as you gain experience.

➥ Start your project with a clean workbench. Put all tools in their place, either in a toolbox or on hooks on the wall in front of your workbench. Brush off stray bits of solder, wire, wood chips, and the like. Keep a clean floor so you won't skip or slide while moving about.

➥ Plan ahead as to what you are going to do for that day or evening and think ahead as you work. Then, when you finish one job, you will be able to go on to the next. You'll be most efficient with the tools and parts you have out.

➥ Don't clutter your work area. Keep articles of clothing off the workbench. Have a special place where you keep your coffee cup or drink glass so you won't spill it while setting it down. Otherwise, your coffee cup will *always* be in the way! If you don't have a special drink place, you might set it down on a screwdriver or pliers, and your project board is not drink-proof.

➥ Arrange your workbench area so that the tools you use are within easy reach. If you work out of a toolbox, keep it to your right if you are right-handed. Just like your coffee cup, have a special place on your bench for your soldering iron so you can instinctively reach for it and not put your hand in your coffee cup. Likewise, you will not be so apt to return your soldering iron to your coffee cup!

➥ As you proceed with your project, remove metal scraps, chips, solder drippings, and the like as often as necessary to keep a clean bench. Metal scraps or sharp tools can scratch or damage the component or equipment on which you are working.

➥ Use the right tool for the job at hand. This is especially important with screwdrivers and tools with jaws such as pliers, sidecutters, and wire cutters. Most importantly, don't use broken tools. This includes screwdrivers, knives, wrenches, wirecutters, vises, and drills. You can easily damage equipment or hurt yourself if a tool slips or breaks. Don't color code your equipment blood red!

➥ Wear protective eye goggles when working with high-speed tools such as drills, bits, braces, sanders, bandsaws, etc. Protect your eyes at these times. Some companies fire employees if they don't obey this simple rule of eye safety (precaution is much cheaper for the company than disability payments).

➥ If you drop a part or tool on the floor, pick it up right away. But be careful where you step. It is so easy to squash an expensive

IC chip! Discard any component such as a resistor or capacitor that looks damaged. Check it before you install it in a circuit (do your troubleshooting early).

➤ If you get hungry while working at your bench, take a few minutes out and don't eat at your bench. A dropped greasy french fry could leave a 1000-ohm short across your circuit board.

➤ Be patient with your progress. Everything won't always work out the way you want it to, but you'll get there.

➤ When you are through for the day or evening, clean up your work space. Put tools back in their place. Give the bench a final brush off. However, leave a component or note out to remind you what you are to do next. Remember, it is psychologically important that you begin your next work period with a clean work area. While you are cleaning, it will give you a chance to wind down gradually. Then, when you come in the next time, you will be able to start off with a smile when you see your clean workbench!

CARE IN USING TOOLS AND EQUIPMENT

It has long been an adage that if you take care of your tools, they will take care of you. This is especially true in any industry where safety is very important. Some companies fire employees if they are observed to be using a tool improperly. Knowing how to use a tool correctly comes with training that teaches you to take care of your tools. Good electronics tools are moderately expensive and they deserve good care if they are to serve you properly and last. You'll do a much better job with tools that are kept in good condition.

In order to keep tools in good operating condition, clean them frequently. Rust can quickly attack expensive tools, so rub them with a lightly oiled cloth to help preserve the metal. In this manner, they will serve you for many years. You will learn to love and admire an old knife or pair of pliers that has served you well for 30 to 50 years of your life.

Iron and steel tools rust easily, so be sure to protect them by cleaning and oiling them frequently. They will then last you a lifetime. Stainless-steel tools do not rust as easily, but once they have been nicked or dropped on a hard surface (such as concrete), rust spots can form when exposed to moisture. Be sure to clean the tools thoroughly and oil them lightly before returning them to your toolbox or their place on the wall or workbench.

Right Tool, Right Job

Another adage you should always remember is to use the right tool for the right job. This is very important. It promotes safety for yourself, others, and the equipment on which you are working. The screwdriver is probably the most important tool you will be using in electronics or other lines of endeavor, so treat it right. Take the time to pick the right size screwdriver head for the job you are about to do. If the screwdriver head is too small, it will probably be damaged and you might as well throw it away. It will always haunt you with its bent head. Also, be sure that you use a flat-head screwdriver for a slot-head screw and a Phillips head (cross) for a Phillips screw. It is so easy to permanently ruin a Phillips-head screw by attempting to use a smaller size blade screwdriver on it. If the screw is tight, the screwdriver will eventually slip and scar the Phillips-head so that when you finally do get the right size Phillips head driver, it will no longer fit the screw properly. Take a few seconds longer and do it right the first time by picking the correct screwdriver for the screw.

Squeeze It Tight

There is always the temptation to use a pair of pliers for a wrench. You can tell yourself that you are good and thorough and that the pliers won't slip from the screw. You do it and the pliers slip and the screw head has been rounded. Don't use a pair of pliers when you should be using a wrench or nut driver.

Wrenching the Wrench

When using a wrench or nut driver to loosen a nut, be sure that you have chosen the correct size. If the wrench is too large for the nut, it might hold at first, but when you apply pressure, it will slip and round off the edges of the nut. After that, you might have trouble using the correct size wrench or nut driver, or the nut will be impossible to remove. Start right (correct) from the start! This adage is true whether working on electronic items, automobiles, or washing machines. Don't slip up!

The Cutting Edge

You'll use a knife often in your electronics work. And because it is so sharp and handy, you'll be tempted to use it to cut copper wire and other things it might appear capable of conquering. But resist the urge! Use it to cut through wire insulation, electrical tape, string, or to scrape insulation off wire. But don't use it as a pair of wire cutters! And remember that the blade of a knife can easily fold up on you, especially when you apply pressure on it in the wrong direction. Be especially cautious not to use a knife in such a manner that you are cutting toward yourself.

Component Handling

Not only should you take care of your tools, but you should also be cautious as to how you handle electronic components. Some of them are delicate. Small parts, such as resistors, capacitors, transistors, solar cells, IC chips, and the like, must be handled carefully because these parts can be damaged by putting too much strain on their wire leads. And be careful not to nick their leads when cutting or working with them, as vibration can later cause the lead to break at that weak point. Be careful when soldering; the copper wire quickly conducts the heat to their body, so use a heatsink to hold the lead so that the heat goes into that and not the component. Heat can easily damage sensitive parts, so use care when soldering them.

Most of these parts can survive a drop to the floor—but not when stepped on—pick them up immediately. A silicon solar cell will be smashed into smithereens if you drop it on concrete or a hard floor, so be especially cautious when handling them. It helps to lay a piece of carpet or rug under your workbench where you will be working. Not only will it help your feet to be standing on a soft surface, but if you drop a delicate part, it will more likely survive the fall and stay where it landed. A piece of rubber pad or foam rubber can also do the job.

Handling of Printed Circuit Boards

Printed circuit boards (PCBs) require special care when handling. Grasp the board from the edges and not the flat sides. This way you avoid getting fingerprints and body oils on the copper foil. Remember, the copper foil on the board is very thin, and you can damage the board by dropping it on the floor, throwing it in your toolbox, laying tools and parts across it. and by applying too much heat from your soldering iron. Too much heat can cause the foil to peel away from the board, and this can cause a circuit malfunction when you try to use it.

A time to be especially careful with the PCB is when you have completed mounting all component parts and are ready to test it. Any foreign material can affect proper operation of the board. Store your neatly assembled board in a clear plastic bag or package so that you can show it off but keep it clean too.

Handling Metal Chassis Circuits

Many electronic circuits are still assembled on metal chassis. The metal chassis and most of its mounted components are more rugged and do not have to be handled as carefully as PCBs. These circuits can sometimes still operate even after parts of the metal chassis may have been bent, scratched, or dented. However, the physical appearance of the chassis will be spoiled, so rough treatment of any electronic or electrical circuit should be avoided.

SAFETY ON THE JOB

Whether working at home on an electronic project or on the job, safety is most important. It should be uppermost in your mind that safety comes first and all other considerations are secondary. If you injure yourself at home while working on a project, it could affect your performance on the job even though your company had no involvement (except for possible lost time from work). However, if you hurt yourself on the job, management is obligated by law to be involved. Even small companies must have insurance on their workers, including workman's compensation, hospitalization, and liability. So your safety habits on the job are very important to your company. But they should be especially important to you. Take time to do a job right.

Shop Safety

Over the years, a number of safe working habits for the electronics professional and beginner have emerged. Your own company might have something similar to these, but they all promote safety on the job.

- Always wear safety goggles when working with an electric grinder to protect your eyes from flying chips of metal or wood. Your eyes are perhaps your most valuable asset!

- Don't hurry while working on a job. This might cause you to cut safety procedures and rules, and it could cost you. Linger your life—don't lose it.

- Unplug your soldering iron and let it cool off before putting it back in your toolbox when the job is completed.

- Keep your work area clear and clean so you and others won't trip, slip, or fall due to parts or materials being on the floor. Push in cabinet drawers so others won't trip or run into them.

➥ Arrange the tools on your workbench so they are easy to reach. Keep them in the same place so you don't spend time looking for them.

➥ Don't use a knife for a screwdriver. They are handy but you can easily cut yourself and possibly break the blade, which can be a dangerous projectile.

➥ Be aware of what might happen if the tool you are using slips, especially a screwdriver. Don't use your legs or parts of your body as a workbench. If the tool slips, you can be hurt.

➥ When using a knife, always cut away from your body so you won't nick or injure yourself.

➥ Be sure your soldering iron cord is free of other objects so you can easily use it and replace it in its holding stand.

➥ When lifting heavy objects with your arms such as a TV set or a heavy-duty power supply, use your legs to provide power. Keep your back straight and at right angles to the floor. Don't bend over, as this force's your back muscles to do more work and can cause a hernia or back sprain. Take time to do it right. Label heavy object's with their weight's such as a power supply in a rack, so you know what load to expect when you take it out of the rack. Use two people for objects over 75 pounds or so. Place this caution on the equipment and call someone beforehand if help is needed.

➥ Use heavy gloves when lifting heavy metal objects that have sharp edges or ends. Avoid cutting yourself on metal burrs that might be present.

➥ Be sure to turn off all power to a piece of equipment, cabinet, or rack. Unplug it to be especially sure so that you won't be surprised by 120 or 240 volts! If you get shocked form any power on the equipment, you might drop it, knock it from the bench, or not be able to let go from the muscle contraction that usually results.

➥ Know where all power outlet switches or breakers are located so you can quickly turn off the power if there is an accident.

➥ If someone does become impaired due to electrical shock, turn the power off and apply cardiopulmonary resuscitation (CPR) techniques.

➥ If you must work on an electrical device with power applied, it is best if you have someone present who knows what to do in case of electrical shock.

➻ Keep your work area well ventilated. Fumes from soldering irons, cleaning chemicals, etching solutions, spray paint, and the like are toxic and could be hazardous to your health.

➻ Wear clothing suitable for your work, especially around rotating machinery such as drill presses, grinders, lathes, and moving belts. Be wary of the tie around your neck, as it could become a noose if it gets entangles in a drill press or lathe. Wear a cap so your hair won't get caught in machinery.

➻ Remove jewelry such as watches and rings so they won't short out a circuit if you should contact them.

➻ Learn CPR. You could save someone's life in an emergency.

Design Safety In

Whether working on your high-fidelity stereo system, personal computer, radio amateur equipment, electronic project, or on the job at work, safety should be designed into your system. While you might think it's the responsibility of the system or component design engineer to build safety features in for the equipment on which you are working, you should always keep safety in mind and be on the alert to prevent accidents. Some good, simple rules follow. You will recognize and follow other rules as you become experienced.

➻ Place the heaviest equipment or components near the floor level when rack-mounting equipment. Label their weights, and call for assistance when removing them.

➻ Do not allow any high voltage or high current points or terminals to be exposed to easy, accidental touch. Place a metal or insulating grid or shield around the terminals. A pair of dropped pliers can produce a fireworks display when they land on a high-current source terminal. Label these points with a red warning sticker stating "DANGER—HIGH VOLTAGE."

➻ Before wiring a metal chassis, remove all metal burrs with a smooth file or sandpaper so all edges are smooth and free of projections that could catch clothing or cut your hands.

➻ Lay out your circuit so that all parts are readily accessible in the event they need replacement. If they are difficult to put in, they will be difficult to take out. You don't want to nick your knuckles or your knickers when working on an electronic device.

➻ Have test points readily accessible so they can be reached easily with a test probe without fear of shorting out a power supply or component when taking measurements. Position the test point in a place where you won't come in accidental contact with a high voltage or current terminal.

➤ Label parts or components with small signs, decals, or stickers if they are physically hot to the touch when operating. The sign not only warns you "HOT" but it also lets you know that it is normal to run hot and that the component (motor, tube, resistor, etc.) is not overheating. Just as you think "buckle up" when getting into your car or truck, you should think "safety" when working in your shop, at home or on the job.

CHAPTER 2

Electron Theory

The *electron* is the smallest electrical particle and has a negative charge. In this chapter, I briefly explain the main components of the tiny atom—the electron, the proton, and in most cases, neutrons.

THE ATOM . . . WHAT IS IT?

Before looking in detail at the electrical components that affect the movement of and help shape the flow of electrons, let's take a close look at the atom itself. Popular theory has it that all matter on earth, in the earth, in our sun, and that of the local planets, stars, and distant galaxies is made up of the basic building block known as the atom. The basic structure of the atom is the same no matter where it is. Every element on earth is made up of atoms, and of the 109 known elements, some were discovered first on the sun before they were found on the earth.

The Hydrogen Atom

A hydrogen atom is the simplest in nature and is shown in Fig. 2-1A, where an electron is shown orbiting around a proton. The electron has a negative electrical charge, while the proton has a positive charge. The two charges balance each other. A basic precept of physics states

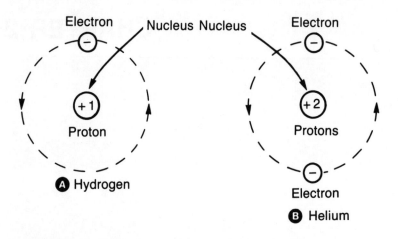

Fig. 2-1. The basic building block of matter—the atom, comprised of at least one electron and one proton. Hydrogen atom (A) and helium atom (B).

that unlike charges attract each other and like charges repel each oth-er. This is also true for the electrical world and the magnetic world (and the human world!). So, according to the unlike-charges attract rule, the electron and proton are attracted to each other. And they would soon contact each other in the atom were it not for the fact that the electron is whirling around the proton at a rate sufficient to balance, or neutralize, the charge attraction. You might have done something similar to this if you have ever tied a rock to a string and whirled it over your head. In the same manner, an earth satellite, or our moon, would go flying off into outer space were it not for the attraction of earth's gravity pulling on that body. Instead, they stay in orbit around the earth. Briefly stated—attraction equals repulsion.

The Helium Atom

The next heaviest atom after hydrogen is helium, shown in Fig. 2-1B. It has two electrons and two protons that balance each other through the "whirling" process. The two positive protons are formed into a tight nucleus with a net positive charge of +2, around which the two electrons orbit.

An Atomic Mystery

The basic building block of the elements continues to build up where protons are added to the nucleus and electrons are added to the orbits. The copper atom, shown in Fig. 2-2, has 29 protons in the nucleus with a total positive charge of +29 and the orbiting electrons

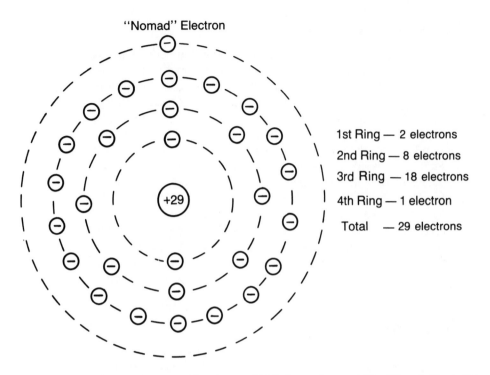

Fig. 2-2. The copper atom has a positive charge of 29 in the nucleus and 29 orbiting electrons. The 29th electron is the "nomad," free to move from atom to atom when "bumped."

have a total negative charge of − 29. The two different charges just balance each other.

As the elements are built up from an atomic weight of 1 (Hydrogen) to 109 (Unnilennium), protons are added to the nucleus and electrons are added to the shells that surround the nucleus. The electron rings, or shells, build up according to a rule that states that each outer ring must be completely filled with electrons before a new ring can begin. The outer ring, however, can never have more than eight electrons before a new ring begins. It can have fewer than eight, but never more.

Let's take a look at the nucleus. In the copper atom, 29 protons are stuffed into a relatively small space. Yet it has been stated that like charges repel. A nucleus containing two or more protons should fly apart and one containing twenty-nine should certainly explode. But it doesn't. Science cannot yet explain this fact completely, but it is theorized that the repelling force in the nucleus is counterbalanced by some attractive force that is not electric in character or nature. We don't yet know what this "super glue" is that defies the attract/repel theory. Some day we may know the answer.

RING AROUND THE ATOM

The building block rule for rings that surround the nucleus states that each ring or shell can contain only a certain number of electrons. The first ring can contain no more than two electrons (such as helium), the second ring no more than eight (such as neon), and the third ring no more than eighteen (such as copper). Thus far we have described rings that held two, eight, and eighteen electrons. But the outer ring never can hold more than the limit of eight as discussed earlier. Once the outer ring builds up to eight, another ring can start with one electron until it builds up to eight. This continues until rings have been built up to contain 2, 8, 18, 32, 32, 15, 2 electrons. The last one given is element No. 109 (Unnilennium), which is the heaviest element discovered so far and exists for just a few microseconds before it decomposes into a lighter element. Of the 109 elements discovered so far by scientists, 17 of them are manmade. These elements have heavier atomic weights, more than the heaviest natural element uranium, which has an atomic number of 92.

The Lonely Electron

Examine Fig. 2-2 again, looking at it closely. The first ring has the required 2 electrons (but no more), the second ring has the required 8 electrons (but no more), and the third ring has the required 18 electrons (but no more) for a total of 28 electrons. However, the copper atom has only one lonely electron in its outer orbit for a total of 29 electrons. It is this lonely, dangling electron that forms the basis of electron movement or current flow in copper and other metals. The understanding of the necessity for this single, lonely electron is profound, because it forms the basis for current flow through anything that is powered by electricity.

The Bumping Electron

In a piece of copper wire, there are trillions and trillions of atoms packed very tightly. Each tiny atom has this single electron whirling madly around in its outer orbit. It is only *lightly* attached to its positive nucleus. In a piece of copper wire without a voltage applied to it and at room temperature, one outer orbital electron can, on occasion, bump into the outer orbital electron of an adjacent atom. The original electron in this outer orbit is "bumped" into another atom as is shown in Fig. 2-3. This electron, in turn, is bumped into another atom, and so the process continues. When an electron enters from the left in Fig. 2-3, the electron bumps an electron from the center atom to the atom on

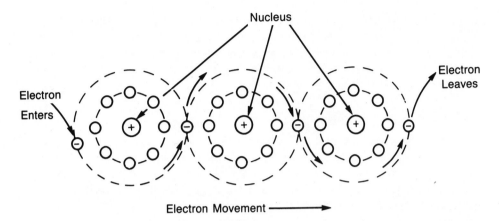

Fig. 2-3. Electron movement through copper atoms in a piece of copper wire. Electron enters from left and is "bumped" to right. (All shells are not shown.)

the right. In a piece of wire without an applied voltage, the movement of the electrons from atom to atom is random, so there is no net electron movement or current flow in the wire in either direction.

When a voltage is applied across the wire, however, with positive polarity to the right as shown in Fig. 2-3, the electron drift is no longer random and the electrons "drift" to the right. Each electron might not travel very far, but there are so many trillions of them in the wire that the apparent drift is at the speed of light. To illustrate this electron drift at the speed of light, Fig. 2-4 shows a tube that is full of ping pong balls. When a ball is pushed in from the left, a ball drops out from the right. Each ball does not move very far but the apparent movement is at the speed of light. The electron movement or motion constitutes a flow of current through the material, be it wire or other material. The speed of the electron movement is near that of light in free space and only slightly slower in a material. For all practical purposes, the flow of current is instantaneous.

The electron ring (or shell) structure of the atom is indeed

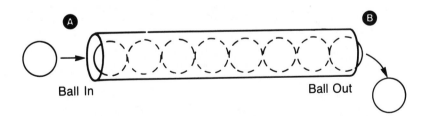

Fig. 2-4. Ping-pong balls in a tube. When one goes in at A, another is pushed out at B. This is likened to electron movement in a wire.

fascinating. Most of the time, the shell build-up follows a rule, but at other times, it doesn't. There is a fine line that separates conductors, semiconductors, and insulators. This division is a relative one, because all three conduct current to some degree, depending on the amount of voltage applied across the material.

METAL CONDUCTORS

Table 2-1 shows the number of electrons in the shells of some common metal conductors. The table is arranged according to the conductivity of the metal, using copper as a reference of 100 percent. Only metal conductors are shown, so a liquid conductor such as mercury or powder conductors such as sodium and potassium are not included.

Note from Table 2-1 that silver is a better conductor than copper by 5 percent, but it costs about 100 times as much. And gold is 70 percent as good a conductor as copper but costs about 5000 times as much. Only King Midas could have his castle wired with gold! At one time in the recent past, home construction used aluminum for wiring. However, because the wire size was the same as copper, many homes caught on fire when the overheated aluminum wire could not properly carry the current. Table 2-1 shows that aluminum is only 61 percent as conductive as copper. That means the wire size for aluminum should be 140 percent of the size of copper in order to be able to carry the same amount of current. Since aluminum is not cost-effective in this case, copper is the most commonly used metal for electrical wiring. Notice that the best conductors have just one electron in the outer shell (silver, copper, gold), where the poorer conductors have two or three electrons (tungsten, nickel, iron).

Table 2-1. Electrons in Shells of Metal Conductors.

Metal Conductor	Relative % Conductivity	Electrons/Shell Total	Electrons/Outer Shell
Silver, Ag	105	47 (2-8-18-18-1)	1
Copper, Cu	100 (Ref)	29 (2-8-18-1)	1
Gold, Au	70.5	79 (2-8-18-32-18-1)	1
Aluminum, Al	61	13 (2-8-3)	3
Tungsten, W	31.2	74 (2-8-18-32-12-2)	2
Nickel, Ni	22.1	28 (2-8-16-2)	2
Iron, Fe	14	26 (2-8-14-2)	2

THE INSULATOR

Conductors are used to maximize current flow. The construction of some atoms is such that electrons do not "flow" from one atom to an adjacent one. These atoms have their outer orbits completely full, or nearly full, with seven or eight electrons, and they hold on very tightly to the outer electrons so that there can be no movement from atom to atom as in a conductor. Materials that have these atomic properties are insulators to electrical conduction.

Table 2-2 shows three elements that have eight electrons in their outer shell. These elements are inert and have no electrical activity. They are neon, argon, and xenon, all gases that will not conduct an electrical current. The same applies to air. Practical insulators are compounds such as porcelain, bakelite, rubber, teflon, glass, and mica. These materials will not conduct unless a very high voltage is applied across the material to pull the electrons loose from the outer ring of the atom. If this occurs, the material can break down. For example, when lightning strikes, the air is broken down. The lightning path conducts hundreds of thousands of amperes of current from cloud-to-cloud or cloud-to-ground. The same thing occurs when you shuffle across a carpet and touch a doorknob. The slight shock you get is the breakdown of the air between your hand and the doorknob. You accumulate static electricity by walking across the carpet, and when you touch the knob, about 10,000 volts goes from you to the doorknob. However, there is little current involved, so all you feel is a sharp, slight shock and you see a short, blue flash.

THE SEMICONDUCTOR

Nature is always full of rules, and as we know, there are always exceptions to the rules. Some materials, such as silicon or germanium,

Table 2-2. Electrons in Outer Shells of Insulators.

Insulators	Relative Conductivity	Electrons Total	Electrons per Shell	Electrons Outer Shell
Neon, Ne	Extremely Poor	10	2-8	8
Argon, A	Extremely Poor	18	2-8-8	8
Xenon, Xe	Extremely Poor	54	2-8-18-18-8	8

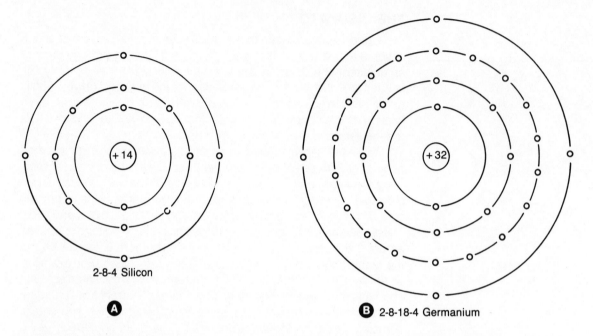

Fig. 2-5. Semiconductor elements with their outer rings half filled with four electrons. Silicon atom (A) and germanium atom (B).

are right in the middle of the muddle—they have too many electrons in the outer ring (4) to be a good conductor and they don't have enough electrons (only 4) to be a good insulator. Figure 2-5 shows these atoms. These elements, or materials, are called *semiconductors* because they act halfway as conductors and halfway as insulators.

The number of electrons associated with silicon, germanium, and carbon are shown in Table 2-3. Note that all of these have 4 electrons in the outer shell. Silicon has desirable properties over the other materials and is used almost exclusively in the manufacture of solid-

Table 2-3. Electrons in Outer Shells of Semiconductor Materials.

Semiconductor	Relative Conductivity	Total Electrons/Shell	Outer Electrons in Shell
Carbon, C	Poor/Good	6 (2-4)	4
Silicon, Si	Poor/Good	14 (2-8-4)	4
Germanium, Ge	Poor/Good	32 (2-8-18-4)	4

state devices such as transistors, IC chips, charge-coupled devices (CCD), diodes, zener diodes, photodiodes, photo transistors, and so on.

THE DOPED CRYSTAL

Pure (intrinsic) silicon or germanium exhibit characteristics closer to that of an insulator rather than a semiconductor. As such, they are useless for transistor work. For them to work properly as a useful semiconductor, certain impurities are added during manufacture in carefully controlled amounts (parts in 10 billion). The ''impurities'' (which in themselves are very pure) create two types of materials, one with an excess of electrons in the crystal lattice created, and one with a shortage of electrons in the crystal lattice (outer shell). Adding impurities to the silicon or germanium is known as *doping* the material.

Doner Elements

The manufacture of semiconductor materials is one of high technology and this is what caused the real breakthrough in the transistor industry throughout the world. The atoms shown in Table 2-4 have 5 electrons in the outer shells. These *pentavalent* (donor) atoms provide, or donate, an extra electron to the 4 in the outer shell of the semiconductor with the 5 of the dopant. Material doped with a donor impurity has an excess of one electron in the crystal lattice structure and is known as a negative or n-type material. However, the net charge of the n-type material is still neutral because the total number of electrons is equal to the total number of protons in the nucleus.

Acceptor Elements

The ''impurities'' shown in Table 2-5 have atoms with only 3 electrons in the outer shell and are called *trivalent* elements. When one

Table 2-4. Electrons in Outer Shells of Donor Impurities.

Donor Impurities	Total Electrons	Electrons in Outer Shell
Nitrogen, N	7 (2-5)	5
Phosphorus, P	15 (2-8-5)	5
Arsenic, As	33 (2-8-18-5)	5
Antimony, Sb	51 (2-8-18-18-5)	5

Table 2-5. Electrons in Outer Shells of Acceptor Impurities.

Acceptor Impurities	Total Electrons	Electrons in Outer Shell
Boron, B	5 (2-3)	3
Aluminum, Al	13 (2-8-3)	3
Gallium, Ga	31 (2-8-18-3)	3
Indium, In	49 (2-8-18-18-3)	3

of these elements is mixed with an intrinsic material, they lock into the crystal lattice structure such that there is a "hole" left in the structure. This hole represents an incomplete bond and exhibits a positive attraction (a missing electron). The bond is completed and forms a stable eight-electron outer shell structure when a free electron is attracted to and is accepted into the bond. This material is thus said to be an *acceptor*. The atom that released the electron becomes positively charged. The net charge of the material is still neutral as the total number of protons equals the total number of electrons. The intrinsic material doped with a trivalent impurity (acceptor) is referred to as a positive or P-type material because the shortage of electrons in the material will attract any free electron.

CHAPTER 3

Electronic Components

Electronic components are the building blocks of all types of electronic circuits that are built or made on a circuit board, metal chassis, plastic frame, or hybrid board. In this chapter, I explain what these components are and how to know and understand what they do. An electronic component is like a brick that goes into making a house or building. There might be many or few bricks, but they are all crucial to the proper stability of the structure.

TYPES OF ELECTRONIC COMPONENTS

Electronic components, such as the resistor or capacitor, are called electronic because they affect the movement of electrons and help shape current flow in a circuit. When first getting started in electronics, all the components look alike. All resistors look alike, and the resistor looks a diode, and the diode looks like a small tubular capacitor, and the capacitor looks like it could be a small inductor, and the inductor looks like it could be an LED (light-emitting diode). How does one avoid much of the confusion and separate out these various devices so things begin to make sense?

Fortunately, there is help after all. It does take a bit of time and understanding to become familiar with all the devices, which is the focus of this chapter. Just as it takes a lot of time, understanding, skill, and experience to learn to change plugs in an auto engine and time it properly, so the learning of electronics takes time, study, and devotion. But the rewards are amazing!

Discrete Components

Before the integrated circuit (IC) was invented (Jack S. Kilby, Texas Instruments, 1959), there was no such a term as a *discrete component*. The term was not necessary, as all components were already unique. A resistor was a component you could pick up individually as with a capacitor or a transistor. They were discrete, not hidden to the eye in a package. Then the IC was invented, now containing thousands of transistors, resistors, diodes, and wire connections. When you pick up an IC, you pick up a whole world of components that are now not discrete to the eye or the hand. However, because you can pick up an IC, it is also listed along with other discrete components!

Passive and Active Components

The basic building block of all the electronics world and its marvels is made up of the following electronic components:

➻ **Passive Components**
Resistors
Capacitors
Inductors (coils)
Semiconductor or vacuum tube diodes

➻ **Active Components**
Electron, or vacuum tubes
Transistors
Integrated circuits

➻ **Hybrid Circuits**
Circuits formed from components of the above groups but packaged to perform a specific function (an early predecessor of the IC)

The resistor, capacitor, inductor, and semiconductor diode are called *passive* because they do not require an external power supply for their characteristic operation. However, the electron tube, transistor, and IC do require a power source and are therefore *active*. The hybrid circuit may or may not require external power, depending on the components involved. However, if it has a transistor or IC, it is active.

ELECTRICAL PROPERTIES

In order to understand the effect that electronic components (resistors, capacitors, inductors) have on a circuit, you must understand the properties they govern. The two most important parameters are *voltage* and *current*.

Voltage

Voltage can be thought of as electrical *pressure*—the higher the pressure, the higher the voltage. The unit of voltage is the *volt,* named in honor of Alessandro Volta (1745-1827), who in 1800 experimented with the *voltaic pile.* The voltaic pile was the forerunner of the voltaic cell which is the forerunner of the wet-cell battery used in your present-day automobile. A battery consists of two or more cells, and in your car battery, there are six 2-volt cells in series to produce 12 volts of direct current. The symbol E is usually used to designate a *source* of voltage, and when a voltage *drop* appears across a device, the symbol V is usually used. However, this distinction is not always strictly adhered to.

Current

Current can be thought of as electrical *quantity*—the greater the current, the greater the electron flow. The unit of current is the *ampere,* named in honor of the French mathematician and physicist Andre Marie Ampere (1775-1836). The symbols I or A are used to represent values for current.

ELECTRONIC COMPONENTS

As complicated as the wizardry of electronics might appear in television, radio, computers, remote planetary probing, astronautics, consumer electronics, medical electronics, etc., there are basically three components that shape the electronics world. These are the resistor, the capacitor, and the inductor. The remaining devices that are of primary importance are electron tubes (formerly called vacuum tubes), solid-state discrete semiconductors, and integrated circuits.

Resistors

Of all the components used in electronic circuits, the resistor is the most common. And, the resistor is probably the oldest component (after the voltaic cell). Although all current-carrying materials have some degree of resistance, the function of the resistor is to "resist" electric

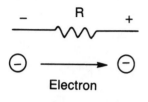

R

– +

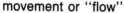

Electron

movement or "flow"

Fig. 3-1. Electrical symbol for a resistor. Electron "flow" produces a voltage drop across the resistor with polarity as shown.

current flow a specific amount by causing electrical friction. That is, it is meant to impede the flow of current to some desired extent, much as you restrict the flow of water through a garden hose by squeezing it.

The electrical symbol for the resistor is shown in Fig. 3-1. When electrons flow into the resistor in the direction indicated, a voltage drop occurs across the resistor with the polarity as shown. References to current, current flow, electron flow, or electron movement all mean the same thing—there is an electrical current due to electron movement.

Resistor Watt Sizes. Low-wattage resistors, such as those used in small electronic devices as radios, hearing aids, pocket calculators, personal computers, and the like are made of a composition paste. This paste is made of carbon (similar to that found in an ordinary lead pencil). Carbon resistors are made in various wattage ratings of ⅛, ¼, ½, 1, and 2 watts. Large-wattage resistors are wire-wound and made in sizes of 5, 10, 25, 50, 150, and 200 watts. The wire-wound resistors are seldom used on printed circuit boards as the copper foil is not large enough to carry the high amount of current required.

The Ohm. The unit of resistance is the *ohm,* named in honor of Georg Simon Ohm (1787-1854), a German physicist who established Ohm's Law. This law expresses the relationship between electric current (I), electromotive force (E), and the resistance (R) of electrical conductors. The law states that a resistance of one ohm exists if one ampere of current is caused to flow through a device when one volt is applied across it. Ten volts will cause ten amperes to flow through the one ohm resistance, and so forth. This relationship is represented as $E = IR$.

The Color Code. When components are large enough, their value and sometimes their use are stamped or printed on the item. However, when the parts are small, a color code is used to indicate the value of the component. In this code, each numerical value is assigned a color as shown:

0 black
1 brown
2 red
3 orange
4 yellow
5 green
6 blue
7 violet
8 gray
9 white

The color code is used frequently for resistors, so it is a good idea to commit it to memory.

Composition (carbon) resistors are marked with three or four color bands to indicate their value. The color code scheme for resistors is shown in Fig. 3-2, which covers all values of resistance. Hold the resistor with the color bands to your left as you read off the value. The first band indicates the first number, the second band the second number, and the third band the number of zeros to be placed after the first two numbers. The fourth band indicates the percentage, plus or minus, that the resistor can be out of tolerance.

As an example, let's say the bands are red, red, and red. The first red is a 2, the second red is a 2, and the third 2 is the number of zeros (00) to add to the 22; the resistor value is 2200 ohms (or 2.2 kilohms). There is no fourth band, so the (default) tolerance of the resistor is ±20 percent.

Take another resistor with coding of brown, black, orange, and silver. Holding the color bands to your left, you see that brown is a 1, black is a 0, and orange is a 3 (for three zeros), so the value is 10,000 ohms (or 10 kilohms). The fourth band is the tolerance margin; silver means ±10 percent.

The third band can also be a multiplier. If it is gold, the first two band numbers should be multiplied by 0.1. If silver, the first two digits are to be multiplied by 0.01. This method lets you identify a resistor that has a value of 6.8 ohms (blue, gray, gold) or 0.68 ohms (blue, gray, silver).

Resistor Tolerance. Let's talk a bit about the fourth band, tolerance. A resistor with a tolerance of ±10 percent means the resistor can have a value 10 percent higher or 10 percent lower and still be called that value of resistance. Take a 100-ohm resistor with a silver fourth band. The resistor can have a value as high as 110 ohms (100 + 10) or as low as 90 ohms (100 − 10) and still qualify as a 100-ohm resistor. If the resistor had a gold fourth band it could have a value

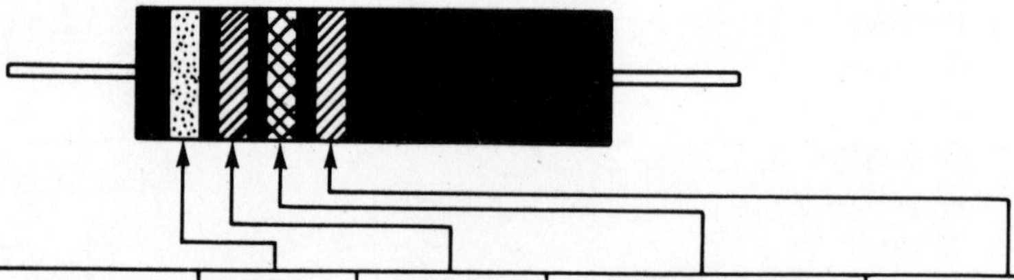

COLOR	FIRST BAND	SECOND BAND	THIRD BAND (MULTIPLIER)	FOURTH BAND
BLACK	0	0	1	—
BROWN	1	1	10	—
RED	2	2	100	—
ORANGE	3	3	1000	—
YELLOW	4	4	10,000	—
GREEN	5	5	100,000	—
BLUE	6	6	1 000,000	—
VIOLET	7	7	10,000,000	—
GRAY	8	8	100,000,000	—
WHITE	9	9	1,000,000,000	—
GOLD	—	—	.1	5
SILVER	—	—	.01	10
NONE	—	—		20

Fig. 3-2. The resistor color code.

from 95 to 105 ohms. The smaller the tolerance, the more precise and the more expensive the resistor.

When two metal plates are placed near but not touching each other, they form a capacitor, or condenser. An electrical "stress," or voltage, can be impressed across the two plates. When a capacitor is charged from a direct current source, the capacitor is said to be fully charged after a certain amount of time. The negative plate of the capacitor has many electrons stored on it, while the positive plate has an absence of electrons (it is positive). A depiction of a charged capacitor is shown in Fig. 3-3. A large capacitor can store many electrons on its negative plate.

The Farad. The unit of capacitance is the *farad*, which is a

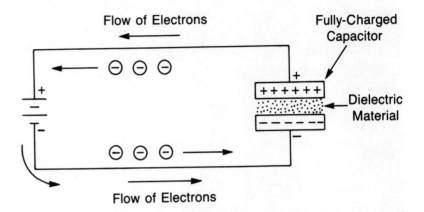

Fig. 3-3. A fully charged capacitor has an abundance of electrons on its negative plate, while the positive plate has lost most of its electrons.

measure of its electrical size. Most capacitors have a value in the microfarad or picofarad range. Until computer power supplies came along (which require great amounts of filtering in their operation), there were no capacitors as large as 1 farad. Radio textbooks as late as the 1950s still described a 1-farad capacitor as being as large as a building or a square city block! However, with the improvement in capacitor manufacturing technology and lower operating voltages, the average small-sized computer might contain a total capacitance of 1 to 5 farads! These large capacitors are necessary to provide a constant operating voltage and actually act as a battery when a "glitch" comes along the power line. Today's capacitors are much smaller than this.

The capacitor retains a charge between the two metal plates that are separated by a *dielectric*, which is a non-conducting material much like an insulator. Capacitors are made using one of the following as a dielectric:

➥ Air
➥ Vacuum
➥ Paper or plastic
➥ Ceramic
➥ Electrolytic
➥ Mica or glass

Some large value capacitors used in personal computers are shown in Fig. 3-4. As described earlier, some of these capacitors can be as large as 500,000 microfarads (0.5 farad). The usual case is that the higher the operating voltage, the higher the value of the capacitor

Fig. 3-4. Large value capacitor.

for a given size. Thus, a large value capacitor, such as 500,000 microfarads, might have an operating voltage of 50 volts dc.

Inductors

The inductor is made of a number of turns of coiled copper wire and is the electrical opposite of the capacitor. The capacitor stores energy in the electrical stress between the two metal plates. The inductor stores energy in a *changing magnetic field* surrounding the inductor. The capacitor can hold an electrical charge for many months if it is made of good dielectric material, meaning the electrons can't sneak back to the positive plate to neutralize its charge. The inductor, however, has a magnetic field generated only when the current through its windings is increasing or decreasing. When the current through an inductor is constant (not increasing or decreasing), the magnetic field collapses and it is no longer functional (only its dc resistance value matters at this time). The schematic for the inductor is shown in Fig. 3-5.

The inductor has a dc *resistance* and an ac *reactance*. The dc resistance is due to the ohmic resistance of the wire in the inductor and the ac reactance is due to the opposition to the flow of current through the inductor. There is a continual interchange of energy between the source of the charging ac voltage and the magnetic field of the inductor. The net result is the magnetic field of the inductor returns

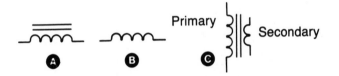

Fig. 3-5. Types of inductors. Iron core is shown in (A), an air core in (B), and a transformer with primary and secondary windings in (C).

as much energy to the source as it draws from the source. Therefore, there is no power consumed by the inductor.

Reactance comes from how the inductor *reacts* to the ac current passing through it. The ac inductive reactance can be as high as ten times that of the dc resistance of the winding. This high reactance limits the current passing through the inductor. The fluorescent light fixture is a good example of where a "ballast" (inductor) is used to limit the amount of current flowing through the fluorescent lamp after it lights. If the ballast has a few turns shorted, it will not produce enough reactance, and excessive current can flow, burning out the lamp.

The Henry. The unit of inductance is the *henry,* named in honor of Joseph Henry (1797-1878), an American physicist. He became famous for his fundamental discoveries in electromagnetism and for his work in improving the electromagnet. To understand the unit of the henry, two coils have a mutual inductance of one henry when a current change of one ampere per second in one coil causes a force (voltage) of one volt to be induced in the other coil.

Electron Tubes

The electron tube is used in a number of commercial, industrial, hobby, appliance, and military products. Previously, it was known as the vacuum tube, and before that it was known as the valve because it in effect turned a stream of electrons on and off. It all started with the Edison Effect.

The Edison Effect. Not only did Thomas Alva Edison invent the electric light bulb in October, 1879, but he also came close to inventing the vacuum tube. In 1883, Edison discovered that if a metal plate was sealed inside the bulb of an incandescent lamp that was operating on direct current, an ammeter placed in the plate circuit would measure a small amount of current. Current would flow when the plate was connected to the positive end of the filament, but none would flow when the plate was connected to the negative end. He also noted that the negative end of the filament wasted away faster than the positive end. The current that Edison observed was caused by electrons emitted

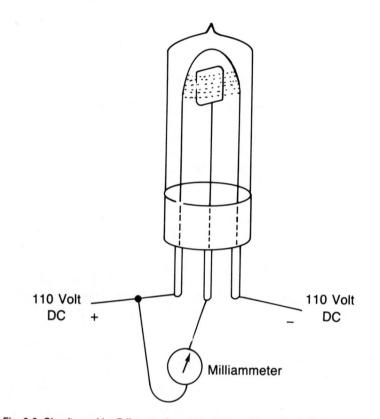

110 Volt
DC   +

110 Volt
DC   –

Milliammeter

Fig. 3-6. Circuit used by Edison to show current flowed from the bulb filament to the plate. This was later called the "Edison effect."

from the lamp filament and attracted to the positive plate. See Fig. 3-6.

Edison did not connect the two phenomena in any way but he did record it in his notebook and it became scientific knowledge. Edison's experiment attracted but little attention for many years, although it involved the fundamental principle underlying the operation of almost every type of vacuum tube known today. That includes the diode, the triode, the tetrode, the pentode, and other complex vacuum tubes that have come along in the century since his discovery of the "Edison effect."

The Fleming Valve. It was but a step from the Edison effect experiment to the Fleming valve experiment. In 1904, John A. Fleming, a British scientist, built the Fleming thermionic valve. This vacuum diode tube could detect radio signals and was called a *valve* because it allowed current to pass in only one direction. This diode was on its way to become the triode.

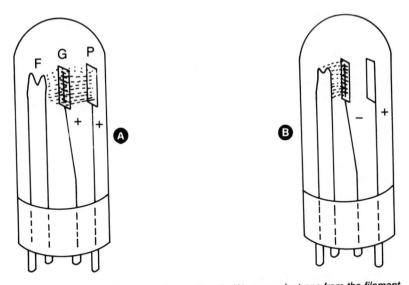

Fig. 3-7. The Lee De Forest triode amplifier. In (A), many electrons from the filament F reach the plate P when the grid G is positive. In (B), the grid is negative and little current flows to the plate.

Triodes. The *triode* was earlier called the three-electrode vacuum-tube detector. The triode tube or *audion* was invented in 1907 by the American inventor Lee De Forest. A wire grid was inserted between the filament and plate of a vacuum tube that controlled the amount of current that reached the positive plate. A small change in the voltage applied to the grid caused a large change in the amount of current that flowed to the plate. The tube thus became an amplifier. The "valve" action is shown in Fig. 3-7A. When the grid wire (G) is positive, it lets most of the electrons from the filament (F) reach the plate (P). However, in Fig. 3-7B, when the grid is negative, it repels most of the electrons from the filament and few of them reach the positive plate.

Modern-day electronics has developed from experiments made in the 1800s with electric currents flowing through glass tubes. All of this work was leading to the marvelous invention of the transistor.

SEMICONDUCTOR DEVICES

Semiconductor materials have been with us for over a century, but we didn't understand enough about them to apply their properties to what was to become the electronics industry.

Diodes. It was around 1880 when it was first observed that current would flow in only one direction through a junction of selenium and a conductor. Yet we had to wait until about 1930 when the commercial application of the selenium rectifier put the discovery to industrial use.

The past always has a way of predicting the future, and this is especially true for the crystal detector used in the early 1900s when radio was in its infancy. The *crystal detector* consisted of a lump of semiconductor material such as galena (lead sulfide), silicon, or silicon carbide that was in contact with a *catswhisker.* The catswhisker was a spring mechanism that could be moved around on the crystal until a very sensitive spot was found that provided the most signal output from the detector into the earphones. It was always necessary to fiddle with the catswhisker to find the most sensitive spot for maximum headset volume, and it was easily disrupted.

The crystal detector became obsolete when the vacuum tube diode was perfected. However, as radio-electronic applications went to higher and higher frequencies, the vacuum tube diode did not have good performance. Crystal diode research returned to make it a better performer at higher frequencies. During World War II, silicon and germanium crystal diodes were used extensively as mixers and detectors in radar receivers and other ultra-high-frequency (UHF) applications. The new crystal diodes were good up to 10,000 MHz and were permanently adjusted to their most sensitive spots. These diodes are still used in modern police radar detectors and radar sets operating up to 25,000 MHz.

Transistors. Before the late 1940s, all electronic equipment used vacuum tubes that consumed large amounts of power. The filaments that were heated to the point where they emitted electrons consumed several watts of power. All electronic equipment was large and bulky, consumed power, and got hot.

The vacuum-tube diode and the semiconductor diode just described are very similar in characteristics. Scientists wondered if it would be possible to add a third element to a crystal diode and get it to amplify the same way a vacuum triode does. Researchers at the Bell Telephone Laboratories set out to see if it could be done. In June, 1948, it was announced by Bell Laboratories that three American scientists—John Bardeen, Walter Brattain, and William Shockley—had invented the *point-contact transistor.* This was the first solid-state amplifier. It was called the transistor and there was exactly one of them in the world. For their work, the three PhDs received the 1956 Nobel price in physics, a real accomplishment. The world would never be the same.

Though miraculous, the first transistors had their drawbacks. The point-contact transistors produced a lot of noise, they had serious frequency limitations, they were unreliable, and it was not possible to produce two alike with similar characteristics. Was there to be an early demise of the transistor?

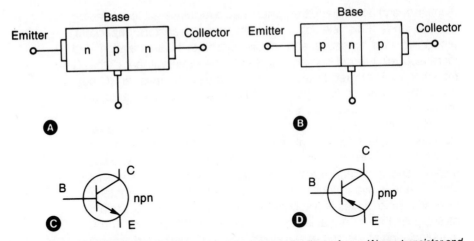

Fig. 3-8. The three transistor regions, showing emitter, base, and collector for an (A) npn transistor and (B) pnp transistor. The schematic symbol for an npn is shown in (C) and for a pnp in (D).

In 1951 Dr. Shockley invented the first *junction transistor*. It was a breakthrough that changed all the rules, because it did everything that it needed to do. The transistor's impact on the electronics industry has been enormous, and there is now a multi-billion-dollar trade throughout the world. The transistor has led to all sorts of discoveries and inventions in other electronic fields such as the integrated circuit, the light-emitting diode, the optoelectronic field, the hand-held calculator, and the microprocessor and microcomputer.

Bipolar means that two kinds of current are effective in the operation of the transistor. Electrons (which are negative) make up part of the current movement and holes (the absence of an electron creates a "hole," which is positive) make up the rest of the current movement or flow in a transistor.

Figure 3-8 shows the three transistor regions that are part of a transistor. The unit is made up of two pn junction diodes that are connected back-to-back, so there are three terminals. The *emitter* emits electrons or holes, the *base* controls the action of the transistor, and the *collector* collects the electrons or holes. Note from Fig. 3-8 that the arrow on the schematic symbol points in the direction of the negative terminal in both the npn and the pnp transistors. This is a handy reminder so that you can easily identify the type transistors used in a schematic diagram without labeling each one.

Integrated Circuits

Man's desire to travel in space helped get the integrated circuit industry started. In the early 1960s, the National Aeronautics and Space

Administration (NASA) and the military had a requirement for replacing the thousands of discrete components (transistors, resistors, capacitors, etc.) used in their space and missile systems. Miniaturization and increased reliability of components led to the development of the IC (affectionately referred to as "chips"). In the early days, an IC cost hundreds of dollars. Now you can buy them at the local radio store for as low as a quarter apiece.

The making of a transistor and an IC is possible only because of manufacturing technology that has evolved to produce such a precision electronic component. A manufacturer first grows a p- or n-type silicon semiconductor material that is several inches long. This crystal is tube-shaped and is two to four inches in diameter. The p-type material is next sliced into pieces of p substrate material and looks like a thin disc as shown in Fig. 3-9A that is about 5 mils thick (.005 inches). Through manufacturing processes known as *diffusion, epitaxial growth, masking,* and *etching,* transistors, diodes, resistors, capacitors, and conductors can be formed on the same chip. Figure 3-9B shows the beginning of this process.

The integrated circuit described above is *monolithic* from the Greek words *monos* meaning single and *lithos* meaning stone. Figure 3-10 shows three components formed on a p substrate—a resistor, a diode, and a transistor. Note how they are all built out of the same material— p and n material on a p substrate. The resistor and diode would be formed using p and n material as shown in Fig. 3-10A and Fig. 3-10B. In a later step, the emitter of the transistor is formed when a second layer of n material is added as shown in Fig. 3-10C.

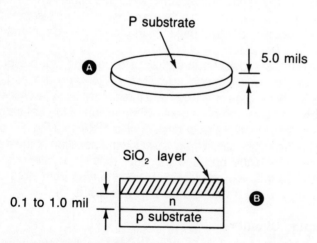

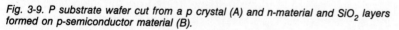

Fig. 3-9. *P substrate wafer cut from a p crystal (A) and n-material and SiO₂ layers formed on p-semiconductor material (B).*

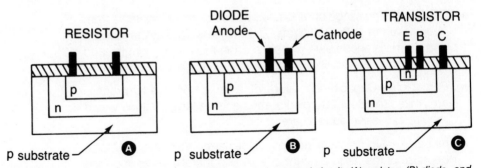

Fig. 3-10. Components that can be used to make up an integrated circuit: (A) resistor, (B) diode, and (C) npn transistor.

Figure 3-11 shows how all the components of Fig. 3-10 are put together to form an integrated circuit, consisting of an input diode connected to an npn transistor that is connected to a resistor. The schematic symbols are shown in Fig. 3-11A and a depiction of a vertical slice through the IC chip is shown in Fig. 3-11B.

Resistors, diodes, and transistors are easy to fabricate on a chip, and almost all ICs use these components. However, inductors and large capacitors are not practical to integrate into chips. Some of the limitations of the components that go into an IC chip are the following.

�» Transistors—Most transistors are npn because pnp are more diffi-cult to produce and usually have less current gain than npn.

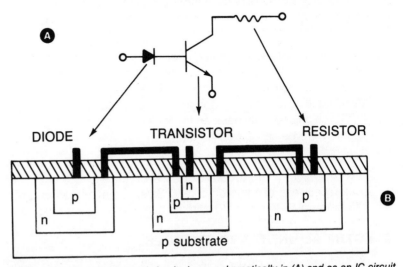

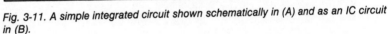

Fig. 3-11. A simple integrated circuit shown schematically in (A) and as an IC circuit in (B).

➤ Resistors—The range of resistance values of an IC resistor is from about 20 ohms to 25,000 ohms. The p layer is used as the resistor, and the resistance value depends on the resistivity, length, and thickness of the layer. Higher values would require too much space.

➤ Capacitors—Capacitance values obtainable from the IC process are limited to about 200 picofarads because larger values would require excessive surface area.

➤ Inductors—No practical inductor values are obtainable in IC form because of the large amount of wire and therefore size required to form an inductor.

An IC can have a few components to many thousands of components per chip. Let's take a look at some numbers.

➤ A *small-scale integration* (SSI) chip has just a few components to form a complete circuit, usually less than 12 integrated components.

➤ A *medium-scale integration* (MSI) chip has from 12 to 100 integrated components.

➤ A *large-scale integration* (LSI) chip has more than 100 integrated components. LSI is defined as the interconnection of 100 or more circuits of logic-gate complexity.

➤ The *very large scale integration* (VLSI) chip has a rather amazing chip density with hundreds of thousands of transistors on a single chip. Current VLSI chips have approximately 90,000 gates and use 275,000 transistors, and this number continues to increase.

➤ The *very high speed integrated circuit* (VHSIC) chip has a goal of 34.7 million transistors on a chip. The goal is not only operational speed but to improve system performance, reduce weight, improve maintainability, reduce cost, and reduce power consumption. Some VHSIC chips presently have an input clock frequency of 32 MHz. Records are being broken regularly.

➤ The *ultralarge scale integrated circuit* (ULSIC) is an IC chip that will have 4 million or more transistors and/or linewidths (traces that conduct current) of 1 micron or less. Sales are expected to reach $35 billion in 1996 with a 70 percent average annual growth rate.

ELECTRICAL UNITS AND SYMBOLS

A summary of the electrical qualities and units is shown in Table 3-1. The names of all the units, such as voltage, current, and frequency

Table 3-1. Electrical Symbols and Units.

Property	Value Designation	Unit	Symbol
Voltage	E or V	volt	V
Current	I or A	ampere	A
Resistance	R	ohm	Ω
Capacitance	C	farad	F
Inductance	L	henry	H
Frequency	f	hertz	Hz
Conductance	G	siemens	S
Power	P	watt	W

use lower case letters, while the symbol for the unit is capitalized, such as V for volt, A for ampere, and H for henry. An exception is Hz, which is the symbol for hertz (cycles per second). For this you would write 60 hertz or 60 Hz. Hertz is both singular and plural.

Multiple Component Designations

When an electronic circuit contains more than one component of the same type, descriptors are used to keep track of them. R1 is the first resistor in a circuit. R2 is the second resistor, and this pattern usually follows from left to right on the schematic. C1 is first capacitor, C3 the third capacitor in a circuit, etc. In the same manner, L1 and L2 describe inductors, Q1 and Q2 identify transistors, U1 and U2 identify ICs, V1 and V2 would describe different voltage levels, and I1 and I2 would describe two currents in a circuit.

Multiplying Factors and Prefixes

When using the units shown in Table 3-1, sometimes the quantities are too large or too small. A 1-farad capacitor is much too large and would not be used. However, one millionth of a farad is more practical in a circuit, and this value is a microfarad. On the other hand, a resistor of 1 ohm is seldom used. But a resistor of 1 million ohms is fairly common in practice (one megohm resistors). Table 3-2 shows the multiplying factors, their prefixes, and their symbols for numbers from 10^{18} to 10^{-18}. Note that the symbols for numbers above 1 million are capitalized (M, G, T, etc.) while for 1,000 and below, small letters are

Table 3-2. Multiplying Factors and Prefixes.

Numerical Value	Multiple	Prefix	Symbol	Common Examples
1,000,000,000,000,000,000	10^{18}	exa-	E	
1,000,000,000,000,000	10^{15}	peta-	P	
1,000,000,000,000	10^{12}	tera-	T	
1,000,000,000	10^{9}	giga-	G	1 = GHz frequency
1,000,000	10^{6}	mega-	M	1 = MΩresistor
1,000	10^{3}	kilo-	k	1 = kHz frequency
100	10^{2}	hecto-	h	
10	10	deka-	dk	
1	1	unit		1m
0.1	10^{-1}	deci-	d	1dm
0.01	10^{-2}	centi-	c	1cm
0.001	10^{-3}	milli-	m	1 = mA current
0.000001	10^{-6}	micro-	$1\mu s$	
0.000000001	10^{-9}	nano-	n	
0.000000000001	10^{-12}	pico-	p	1nm
0.000000000000001	10^{-15}	femto-	f	1 = pf capacitor
0.000000000000000001	10^{-18}	atto-	a	

used (k, h, dk, etc.). These prefixes are taken from the International System of Units (SI) and are used by most countries in the world. Remember to always use them correctly so that when they precede a unit that is capitalized, such as V for volt, write kV for 1,000 volts, and 10 mA for 0.01 amperes. Remember to always use a small m for milli (mA, etc.) and the Greek letter mu (μ) for micro (10^{-6} or one-millionth).

CHAPTER 4

Electronic Diagrams

A schematic diagram is a road map of an electronic circuit, such as a television, a high-fidelity stereo amplifier, or a lamp dimmer. The symbols for the various "roadmarks" along the way are international in nature, so they are recognized by just about any electronic engineer or technician in the world. The stereo receiver that was designed and manufactured in the Far East can be easily maintained in the United States because the circuit diagrams are universal. The symbol for an npn transistor in Japan is the same in the United States, and a electronic product manufactured in the United States can be repaired in Japan.

BLOCK DIAGRAMS

A block diagram is a representation of an electronic device or major appliance that shows how several groups of circuits interact without depiction of all the discrete components. A radar transmitter is shown in Fig. 4-1. Note from this figure that there are no electronic components inside the blocks, just an indication of what the blocks do to make the radar work. In the block diagram labeled *power supply*, there is no indication as to what is within the box, but we do know that the box supplies power to all of the other transmitter component blocks. The diagram shows you what it takes to make up *this* radar set, but others might be slightly different. However, they must all have the same

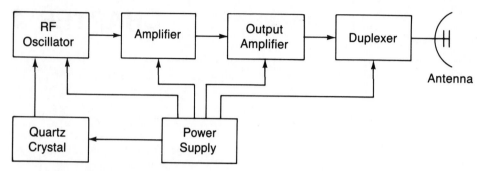

Fig. 4-1. Block diagram for a radar transmitter.

general building blocks—an oscillator, an amplifier, an antenna, and a power supply.

PICTORIAL DIAGRAMS

A pictorial diagram is a physical representation of the circuit component layout, as you see illustrated in Fig. 4-2, whereas a

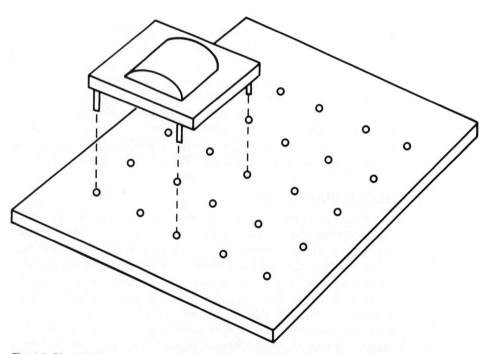

Fig. 4-2. Pictorial diagram showing the placement or mounting of a transformer on a circuit board or chassis.

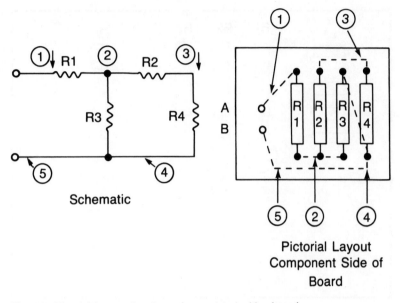

Schematic

Pictorial Layout
Component Side of
Board

Fig. 4-3. Pictorial layout of resistors on a printed wiring board.

schematic diagram is a symbolic representation of an electronic circuit. Pictorial diagrams sometimes show the actual sizes of components as well as placement on a circuit board, but schematic diagrams depict the components in symbol only, with no regard to physical size, placement, or shape.

Figure 4-3 shows a schematic diagram of four resistors and how they would appear when shown in a pictorial layout of an etched circuit board. Note that all resistors are parallel and near each other and that there are no crossover paths. A pictorial diagram is often used on an assembly line to show how the electronic parts are to be mounted or how they are to be connected.

LAYOUT DIAGRAMS

When a pictorial diagram or layout is used to show how parts are layed out on a PCB or chassis, it is called a layout diagram as shown in Fig. 4-4. This diagram shows the relative position, size, and shape of the components as they appear on the board. A drawing such as this is especially useful and necessary when there are many workers on an assembly line that must mount the components in the same manner. A quick glance at the diagram shows that all components have been inserted in the board or chassis. (No wiring is shown.)

43

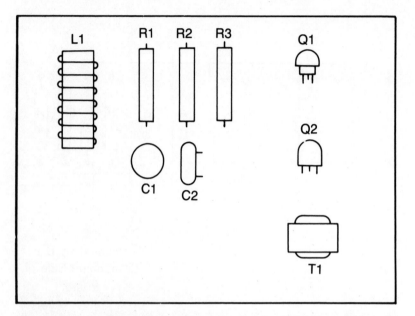

Fig. 4-4. A layout diagram shows how components are arranged on a PC board or chassis. Component positions are exact.

WIRING DIAGRAMS

A wiring diagram is a cross between a circuit diagram and a pictorial diagram because it indicates how parts are connected together. Figure 4-5 shows a simple circuit that connects a flashlight battery, a small bulb, and a switch with the associated wiring. For a PCB, the

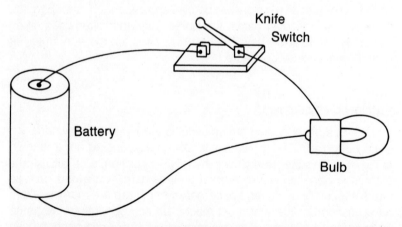

Fig. 4-5. A wiring diagram shows how components are connected together to form a circuit. Component positions are relative.

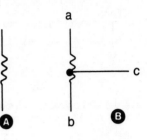

Fig. 4-6. The symbol for a fixed resistor is shown in (A) with two terminals and (B) with three terminals.

"wiring" is etched on the board so that when the parts are inserted and soldered in place, the circuit is completely wired.

SCHEMATIC DIAGRAMS

This section explains some of the symbols that go into making a schematic diagram. We've covered some of these in earlier chapters but let's take a detailed look at some of the more prominent components. The components covered are resistors, capacitors, inductors, transformers, and miscellaneous other important symbols.

Resistors

A "zig-zag" line is used to represent the resistor, and they can be fixed, variable, fusible, or special purpose.

Fixed Resistors. The symbol for the fixed resistor is shown in Fig. 4-6. Figure 4-6A is the standard two-lead fixed resistor, while Fig. 4-6B is a three-lead fixed resistor with different fixed values of resistance available from points a to c and b to c. The value of fixed resistors can be from a few tenths of an ohm to many megohms. The power-handling capability of the discrete resistor can vary from as little as ⅛ watt to many hundreds of watts.

Variable Resistors. Figure 4-7 shows the symbol for a variable

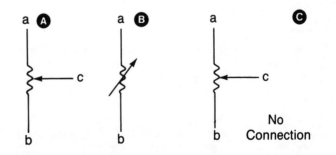

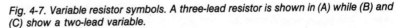

Fig. 4-7. Variable resistor symbols. A three-lead resistor is shown in (A) while (B) and (C) show a two-lead variable.

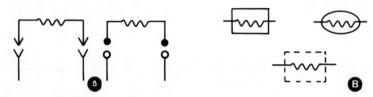

Fig. 4-8. Fusible resistors used to protect electronic circuitry from current overload. Shown in (A) are plug-in and (B) hard-wired symbols.

resistor. Figure 4-7A shows a three-lead variable resistor as is used in the volume control of a radio, with circuit wires connected to points a, b, and c. Figure 4-7B shows a variable resistor with the arrow through the resistor to symbolize that the resistance is variable or adjustable. Figure 4-7C shows a three-terminal variable resistor of which only two terminals are used (a and c). Terminal b is not used. Variable resistors such as those in Fig. 4-7B and C are used as tone controls in radio receivers.

Fusible Resistor. A fusible resistor is a special case of a resistor used to act as a fuse. Such a "resistor" is nominally of a low value of resistance so that it will open after a certain value of current is exceeded. Figure 4-8 shows a variety of symbols used to indicate a fusible resistor. Figure 4-8A shows those resistors which are plugged into the circuit and are easily replaceable, while Fig. 4-8B shows resistors that are soldered into the circuit to offer one-time overload protection. They must be unsoldered to be replaced.

Special-Purpose Resistors. Special-purpose resistors can be made by using special formulas in the composition of the resistors. The resistors can be made to be sensitive and change their value up or down when the voltage, current, temperature, light, or some other quantity varies. The symbols shown in Fig. 4-9 include T for temperature, V for voltage, I for current, and L or lambda (light wavelength) for light-controlled resistors.

Capacitors

The next most used electronic circuit device after the resistor is the capacitor, which stores an electron charge between two plates that

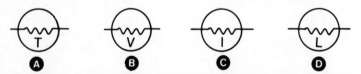

Fig. 4-9. Special-purpose resistors vary their value (up or down) depending on temperature (A), voltage applied (B), current (C), and light applied (D).

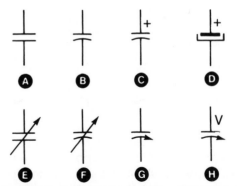

Fig. 4-10. Symbols for capacitors. Fixed values (A) and (B), electrolytic (C) and (D), variable values (E), (F), and (G), and voltage variable (H).

make up the capacitor. Figure 4-10 shows the various symbols for the different types of capacitors available. Capacitors of fixed value are shown in Fig. 4-10A and B, electrolytic capacitors are in Fig. 4-10C and D, variable capacitors are in Fig. 4-10E,F, and G, and a voltage-variable capacitor is shown in Fig. 4-10H. Capacitors work with ac or dc voltages applied, depending on the type of capacitor.

Inductors

An inductor is formed when a piece of copper wire is wound in a circle to form a coil and current is passed through it. The current can be ac or dc and each causes the coil to react differently. The coil is a basic element in radio circuits, relays, solenoids, telephone handsets, loudspeakers, electric trains, the rail gun, and the like.

Air Inductor. The simple air inductor is shown in Fig. 4-11. A fixed value of inductance is shown in Fig. 4-11A. The variable-value inductor in Fig. 4-11B was used in the early days of radio where a slider contact slid along on the copper coil to tune in different radio stations. An adjustable air inductor is shown in Fig. 4-11C, and a tapped, fixed value of inductance is shown in Fig. 4-11D.

Magnetic-Core Inductor. The wire used in an inductor can be wound around an iron core, forming a magnetic-core inductor that has

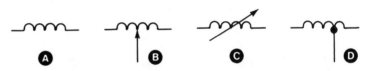

Fig. 4-11. Air inductors formed by coiled turns of wire. Fixed value (A), variable value (B), adjustable (C), and tapped (D).

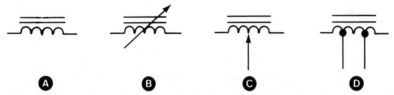

Fig. 4-12. Magnetic-core inductors. Fixed value (A), variable value (B) and (C), and stepped value (D).

much more inductance than the air-core inductors. These magnetic types of inductors are used in power supply filters, audio filters in stereo high-fidelity amplifiers and any application that calls for a large amount of inductance. Figure 4-12 shows the symbols for various types of magnetic-core inductors.

Transformers

A transformer "transforms" a voltage from one level to another (up or down). It also can act as a *reactance transformer,* stepping *impedances* up or down. The output transformer in your home or auto stereo amplifier transforms the output impedance of your amplifier from hundreds of ohms to the four or eight ohms of your loudspeaker.

Air-Core Transformer. As with the air-core inductor, the air-core transformer has a "core" of air and is used primarily in radio-frequency circuits of a radio receiver or transmitter. Figure 4-13 shows the symbol for the air-core transformer with a one-to-one turns ratio in Fig. 4-13A, a step-up transformer in Fig. 4-13B, and a step-down transformer in Fig. 4-13C, and a tapped secondary transformer in Fig. 4-13D.

Iron-Core Transformer. The iron-core transformer is probably the most popularly known transformer in the world. It is used in almost every electronic or electrical device that operates on an electrical voltage, either in the power supply, intermediate amplifier stages, or output stages. This list includes the telephone in your home or office, your transistor radio, microwave oven, TV set, VCR, power line to your home or apartment, and hundreds of other electronic and electrical devices. Figure 4-14 shows the symbols for the various iron-core or ferrite-core transformers.

Fig. 4-13. Air-core transformers. One-to-one coupling (A), step-up (B), step-down (C), and center-tapped (D).

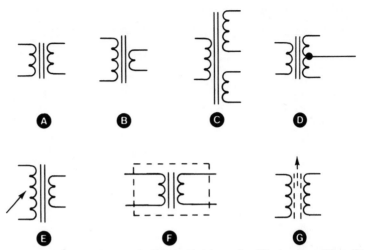

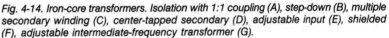

Fig. 4-14. Iron-core transformers. Isolation with 1:1 coupling (A), step-down (B), multiple secondary winding (C), center-tapped secondary (D), adjustable input (E), shielded (F), adjustable intermediate-frequency transformer (G).

Additional Circuit Symbols

There are many additional circuit symbols that are shown in the figures that follow. Figure 4-15 shows the symbols for switches, Fig. 4-16 shows plugs and jacks, Fig. 4-17 shows voltage and current sources, and Fig. 4-18 shows various diodes.

The symbols used for npn and pnp transistors are shown in Fig. 4-19, while optically-coupled isolators are shown in Fig. 4-20. An

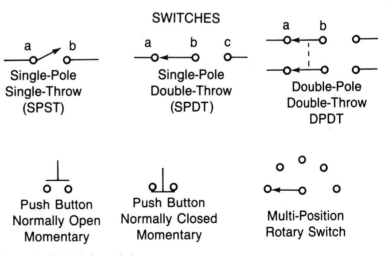

Fig. 4-15. Symbols for switches.

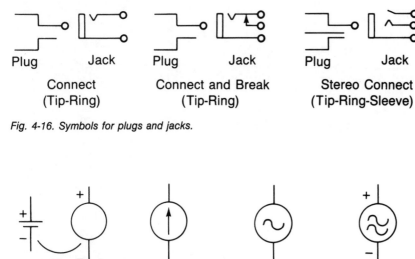

Fig. 4-16. Symbols for plugs and jacks.

Constant Voltage
(Dc)

Constant Current
(Dc)

Ac
Oscillator
(Ac Generator)

Piezoelectric
Sounder
(Audio)

Fig. 4-17. Symbols for voltage and current sources.

Anode
Cathode

Rectifier
(Forward)
Conducting

Light-Emitting
Diode

Photo Diode

Zener
(Breakdown)

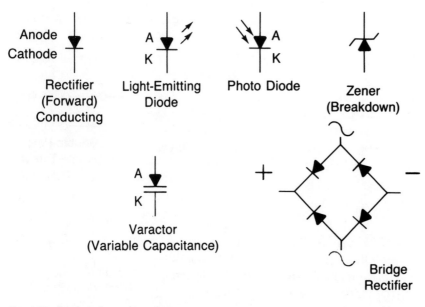

Varactor
(Variable Capacitance)

Bridge
Rectifier

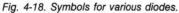

Fig. 4-18. Symbols for various diodes.

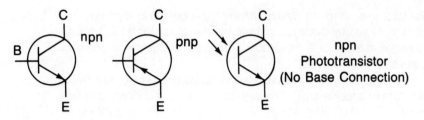

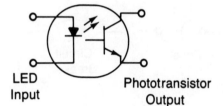

Fig. 4-19. Symbols for transistors.

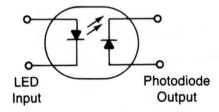

LED Input   Photodiode Output

LED Input   Phototransistor Output

Fig. 4-20. Symbols for optically-coupled isolators.

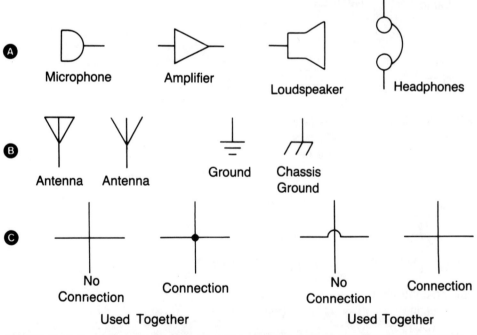

Ⓐ
Microphone   Amplifier   Loudspeaker   Headphones

Ⓑ
Antenna   Antenna   Ground   Chassis Ground

Ⓒ
No Connection   Connection   No Connection   Connection

Used Together   Used Together

Fig. 4-21. Symbols used in audio amplifiers (A), radio transmitters and receivers (B), and circuit connections in (C).

51

optoisolator is a coupling device in which the coupling medium is a light beam. Note that the arrow in both the transistors and LED points *to* the negative terminal, the cathode. The arrow itself is the anode and is positive.

In Fig. 4-21 are the symbols used in audio circuits (A) and radio transmitters or receivers (B). The symbols for circuit connections are shown in Fig. 4-21C. The connection and no-connection symbols to the left in (C) are always used together, while those to the right in (C) are always used together. One or the other is used in a circuit diagram

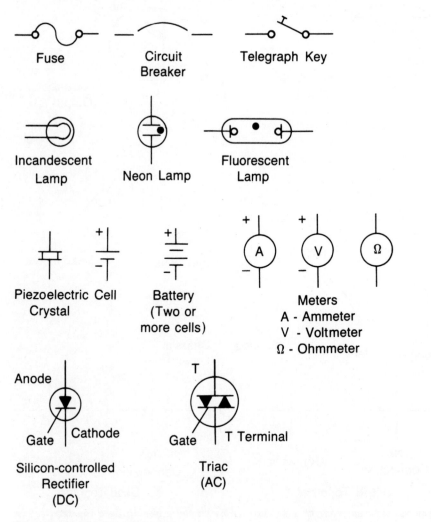

Fuse　　Circuit Breaker　　Telegraph Key

Incandescent Lamp　　Neon Lamp　　Fluorescent Lamp

Piezoelectric Cell Crystal　　Battery (Two or more cells)　　Meters
A - Ammeter
V - Voltmeter
Ω - Ohmmeter

Anode　　Cathode　　Gate
Silicon-controlled Rectifier (DC)

T　　Gate　　T Terminal
Triac (AC)

Fig. 4-22. Additional circuit symbols.

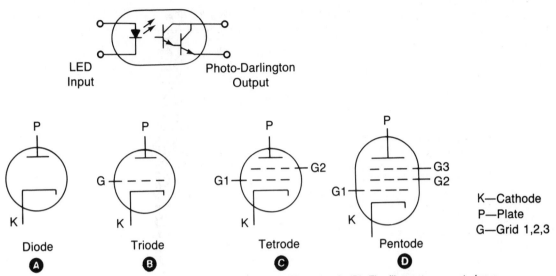

Fig. 4-23. Vacuum tube symbols. Diode (A), triode (B), tetrode (C), pentode (D). The filaments are not shown.

but never one from each set. A glance at a circuit diagram will show you which one is being used.

Additional symbols are shown in Fig. 4-22, while Fig. 4-23 shows the symbols for vacuum tubes. The vacuum tube is used very little today, because its bulky size and high power requirement have been replaced by the transistor and the very nimble and versatile IC chip. However, in some high-power radio transmitters, some television receivers, and special electronic applications, vacuum tubes are still in use.

Understanding Schematics

Figure 4-24 is a schematic diagram for a small-signal linear IC audio amplifier. This diagram is an electronic road map of the circuit that makes up the amplifier.

The input to the circuit is to the left at "Audio Input" and the output is at the right at "Speaker." The circuit operation in schematics usually proceeds from left to right following a logical process. As you begin to understand how the different components operate, you become familiar with how the various diagrams resemble a road map. An electronic current is "traced" from input to output.

Larger electronic circuits use numbers to designate the various components that go to make up the circuit. Refer to Fig. 4-25. All the resistors are designated with numbers, and with the input to the left, the resistor numbers are low to the left and get larger toward the right.

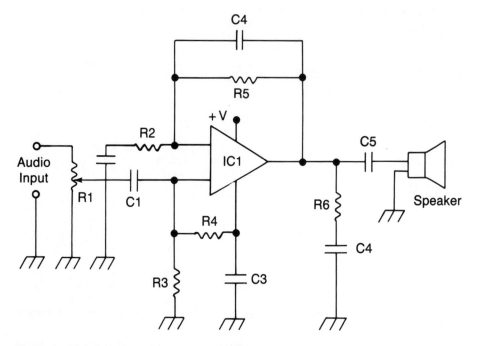

Fig. 4-24. Schematic diagram for a linear IC amplifier circuit.

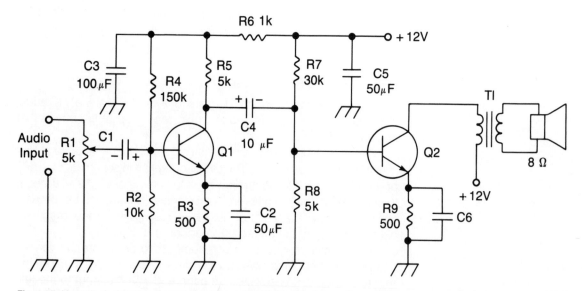

Fig. 4-25. Numeric designation of components on a circuit diagram of an audio amplifier.

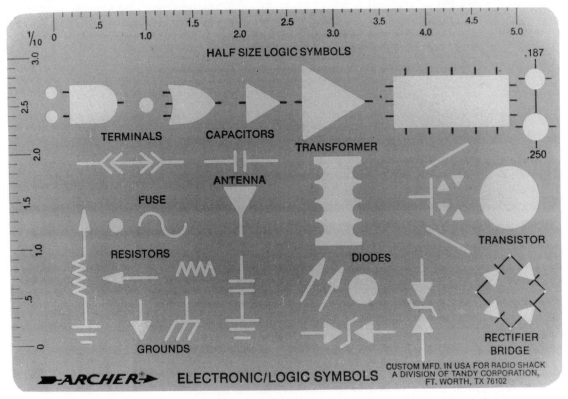

Fig. 4-26. Template for drawing electronic and logic symbols (Courtesy Radio Shack, Division of Tandy Corp).

Similarly, the first capacitor (C1) is to the left, and the last one (C6) is to the right. Likewise, the first transistor is Q1, at the input, and the output transistor is Q2, which drives the output transformer, T1, the first transformer used in the circuit.

When drawing circuit symbols, a more professional-looking circuit diagram will result if you use one of the plastic templates available at various electronic stores. Figure 4-26 shows such a template that illustrates some of the symbols used in electronic and computer logic design. There are easily 10 to 15 different such templates available to cover all aspects of electronics, computer logic, electrical work, programming, communications, etc.

CHAPTER 5

Electronic Tools and Soldering

Electronic equipment fabrication involves two main areas—mechanical assembly and electrical assembly. Mechanical assembly is the simple process of mounting parts on a PCB or placing the board in a chassis or case. The chassis can be further mounted in a cabinet such as your home stereo rack or when you replace your TV chassis in its cabinet. The electrical assembly involved consists of wiring and soldering. This chapter explains some of the hardware, tools, and techniques of electronic assembly.

HARDWARE

The types of hardware used in electronic assembly refer to all the connectors, lugs, terminals, and clamps as well as all the electronics components and wires that make up the assembly or PCB. In other words, two broad categories of parts are those that provide physical support to the circuit or board and those that perform electrical functions in the circuit.

Solderless Connectors. An assortment of solderless connectors is shown in Fig. 5-1. Wires are crimped on with a crimping tool.

Solder Lugs. See Fig. 5-2 for an assortment of lugs for use when soldering to copper wire. The wire is inserted into the small hole in the

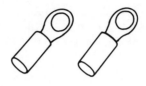

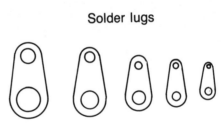

Fig. 5-1. Solderless connectors that are crimped on the wire with pliers or crimping tool.

lug, bent around it, and soldered. The large hole is then connected to a screw terminal, such as a barrier block.

Terminals. Component leads from resistors, capacitors, diodes, transistors, or wires are connected to terminals that are mounted on a terminal strip or barrier block.

Solder lugs

Solder lugs
internal teeth

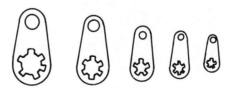

Fig. 5-2. Solder lugs that are used to connect wires to binding posts or screw terminals.

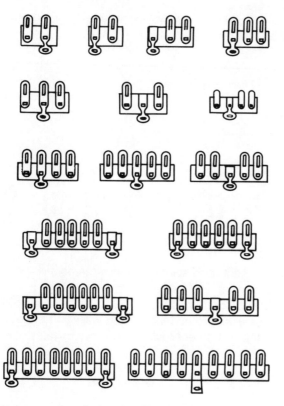

Fig. 5-3. Various types of terminals strips. The support lead can be used to connect to ground.

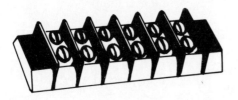

Fig. 5-4. Barrier blocks have screw terminals to which connections are made.

Terminal Strip. Terminal strips are shown in Fig. 5-3. The terminals can be connected to the power supply, main or chassis ground, or any other desired connection. They can have any of a number of terminals. The figure shows strips from one to eight terminals plus ground.

Barrier Blocks. Barrier blocks have screw terminals to which connections are made. Connections to the block have insulated islands separating the adjacent terminals. Wires are connected to the screw terminals directly or by use of a lug terminal. See Fig. 5-4.

TOOLS IN GENERAL

Electronic assembly is considerably easier if the proper tools are available while working on electronic circuits and boards. Tools for mechanical assembly include pliers, screwdrivers, and perhaps a knife. However, for electrical assembly, your soldering iron is the most important tool. There is no doubt that soldering is one of the most important tasks in electronics assembly. A poor soldering job is bound to fail whether it's in a few weeks or a few years, but it will happen and could be very difficult to find once it does occur.

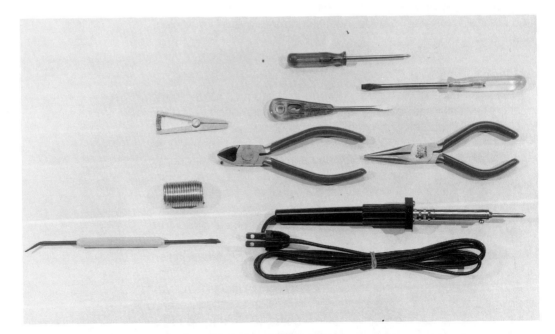

Fig. 5-5. This ten-piece electronic tool set includes screwdrivers, two kinds of pliers, a soldering iron, soldering tool, solder, and heatsink. (Courtesy Radio Shack, Division of Tandy Corp.).

Figure 5-5 shows the tools needed for basic electronic construction, including various screwdrivers, cutters, pliers, and a soldering iron with its accessories. Especially helpful are insulated-handle long-nose pliers with a cutting side for clipping or trimming wire. These can be used for wrapping wires around terminals or for holding components while soldering. These pliers also act as a heatsink when soldering leads to transistors or diodes.

Other pliers frequently used in electronics assembly include round-nose and chain-nose pliers. The round-nose pliers are used as a forming tool to shape wire and component leads around its jaw without damage or stress on the leads.

When removing the insulation from covered wires, a knife can be used, but it could easily nick the wire and possibly weaken it. Wire strippers come in many shapes and sizes. With a "dial" stripper tool, just set the small dial to the wire size you are going to strip and it cleanly removes the insulation. You can then also cut the wire with the tool. Most cover the range of wire gauges 12 to 24 and come with insulated grips.

SOLDERING TOOLS

Soldering tools consist of the irons and their various tips, stands, and other aids for making soldering faster, easier, and more efficient. See Fig. 5-6.

Soldering Irons

The real purpose of a soldering iron is not to melt solder but to heat the junction to be soldered to a high enough temperature to then melt the solder when it is applied. For most electronic work, a small soldering iron of 25 to 50 watts is recommended, as this produces enough heat to do the job yet avoids damaging sensitive components.

Micro Soldering Iron. This small iron provides fast warmup (Fig. 5-7) because it is only 15 watts. It supplies heat to a long-life, slip-on tip that is replaceable. It is especially useful for printed circuits and other microelectronic applications. The iron is available at local stores that stock GC Electronics items.

Temperature-Controlled Soldering Station. A soldering station is shown in Fig. 5-8. The station regulates the temperature over a range of 400 to 900 degrees Fahrenheit, with a pilot light to indicate the rate of power consumption. A "pencil holder" is provided for storage, and the rotary wiping sponge is split to allow both sides of the tip to be cleaned at one pass. The temperature of the iron is controlled by switching the ac voltage at zero-crossing points so that it does not cause radio

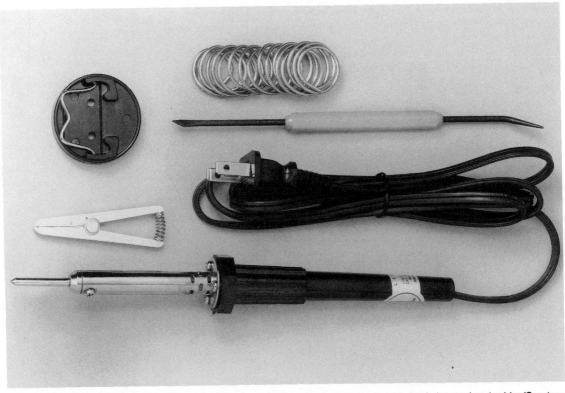

Fig. 5-6. Tools used for soldering include a soldering iron, heatsink clamp, iron stand, soldering helper tool, and solder (Courtesy Radio Shack, Division of Tandy Corp.).

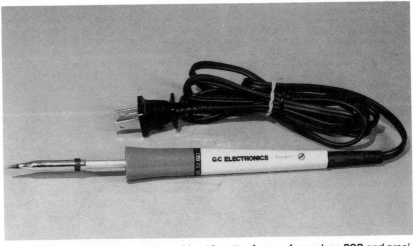

Fig. 5-7. This micro soldering iron provides 15 watts of power for work on PCB and precision electronics (Courtesy GC Electronics).

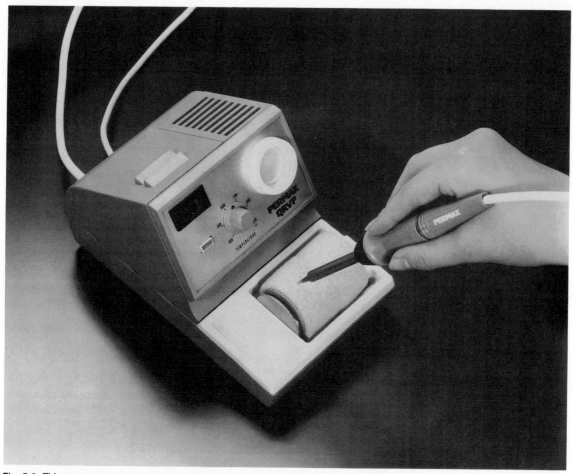

Fig. 5-8. This temperature-controlled soldering station has an adjustable control for the iron over a wide range (Courtesy J.J. Blair Associates, Inc.).

frequency interference. This iron is safe for soldering CMOS ICs that can be easily damaged by static discharge from soldering equipment or other handling. The Permax QRVP 200 soldering iron is available from J. J. Blair Associated, Inc., 10 Church Street, P.O. Box 91, Liberty Corner, NJ 07938-0091.

Tips

The tip shapes seen in Fig. 5-9 include screwdriver, long screwdriver, narrow screwdriver, conical flat, long conical, bent conical, single flat, chisel, pencil cone point, tapered needle, micro spade,

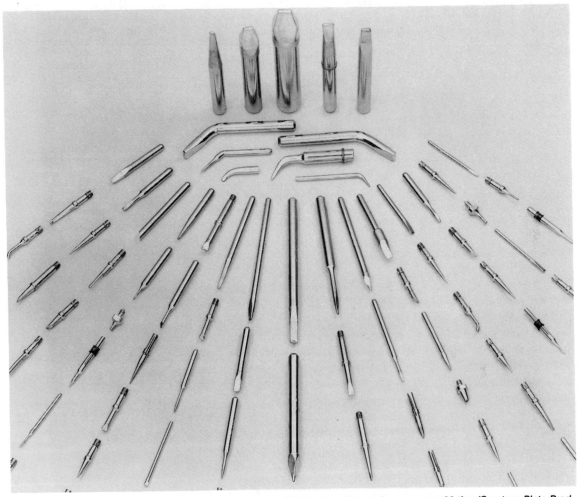

Fig. 5-9. There are many types, sizes, and shapes of soldering iron tips. Shown here are over 80 tips (Courtesy Plato Products, Inc.).

spade, stepped spade, short spade, needle, pyramid, offset chisel, and long taper chisel—over 80 tips are shown.

Surface-mounted devices (SMDs) are the latest ICs available. Figure 5-10 shows an assortment of soldering tips for use with SMDs. The tips are made in two pieces—an adapter that fits the soldering iron to be used and a tip that fits securely over the adapter. The tip has an end shape designed to solder or desolder the specific type of SMD at hand.

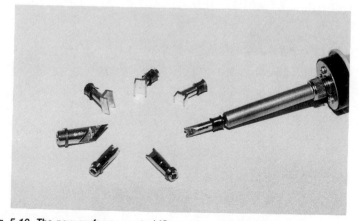

Fig. 5-10. The new surface-mounted ICs can be soldered and desoldered using this special tip that fits over the chip (Courtesy Plato Products, Inc.).

There are two types of soldering irons—those with copper tips and those with iron-plated tips.

Copper Tips. Even though they are called soldering "irons," most soldering iron tips are made of copper for the best heat conduction. Tips must be smooth, and when soldering, you should "tin" (coat with solder) the tip so that heat can be more easily transferred from the tip to the work to be soldered. To tin the tip of the iron, it must be heated long enough to the point where it changes color. Then touch rosin-core solder to it. The solder should flow easily over the tip so it has a silvery look.

When the iron becomes well-used, the copper tip can become pitted as a result of normal wear. The tip should then be thoroughly cleaned. The easiest way is to let the iron cool and clean it with a sponge. It might be necessary to file a copper tip to smooth all the flat sides before you heat and retin it for use. Scaly oxide deposits should also be removed from the tip by rubbing with a 100-grit emery cloth, or use a polishing bar of 80-grit abrasive polyurethane foam stick.

Iron-Plated Tips. Many soldering iron tips are iron-plated to ensure longer life. These tips are usually "pre-tinned" for easy use. Figure 5-11 shows the composition of an iron-plated tip. Iron-plated tips resist pitting from the molten solder that can ruin a copper tip in just a few hours, especially if it sits on a stand without being used. Note that there is an inner plate of nickel for corrosion protection and an outer plating of chromium to immunize the surface from solder. Solder should be applied only to the iron tip and not any other portion of the soldering iron.

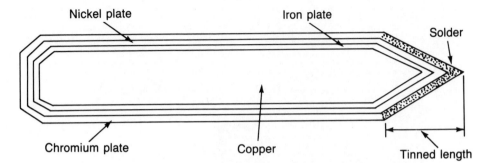

Fig. 5-11. Construction of an iron-plated soldering tip.

The iron-plating on a copper tip is only between 0.003 and 0.030 inches thick but can increase the life of the tip up to 100 times longer over a non-plated tip. This is because iron dissolves much slower (dissolving caused by the molten tin of the solder). Some iron-plated tips have additional working surfaces that serve to protect the tip, such as nickel, silver, tin, or gold. It is then necessary to flood these surfaces with rosin-core solder during the initial heating of the iron, or the platings could burn off and lose their protective capability. A plated tip can remain idle for a long time without damage as long as it remains tinned.

Other Tools and Aids

A number of soldering tools and aids are available for use by all levels of hobbyists and professionals. These are some of the devices and how they are used to make your work in electronics easier and more professional looking.

A small hand drill is shown in Fig. 5-12. This drill is good for drilling holes in PCBs and light materials. Two small bits are kept in the handle for storage.

Sometimes it is helpful to have a third hand when soldering (to hold the solder, the component or board, and the soldering iron). The "helping hands" shown in Fig. 5-13 free up your hands so you can work more carefully and efficiently on your project. The "hands" can be adjusted to any angle to save you hours of frustration.

A special tool used in electronic assembly is shown in Fig. 5-14. This is a balanced, all-metal wire-wrapping tool for making solid wrapped connections. With the wire inserted, just twist it. This tool is much faster than soldering and can be used to strip, wrap, and unwrap 30-gauge single-strand copper wire.

Fig. 5-12. A small hand drill is used for drilling small holes in PCBs and has two bits stored in the handle (Courtesy Radio Shack, Division of Tandy Corp.).

A wire-wrapping kit is shown in Fig. 5-15. This kit comes complete with a PCB, a wire-wrap tool, a roll of wire-wrap wire, two 14-pin DIPs, and two 16-pin DIP plastic IC sockets.

When working with ICs, it is convenient—and wise—to have the proper tools to insert and extract the ICs to prevent damage. Figure 5-16 shows different types of such tools that are widely used by major retailers and electronic service personnel. These include DIP IC extractors and inserters to accommodate all ICs from 8 to 40 pins. The kit shown contains extractors that can handle all large-, medium-, and small-scale (LSI, MSI, and SSI) devices from 8 to 24 pins. The extractor

Fig. 5-13. The "helping hands" project holder is adjustable to almost any position and is especially handy when soldering (Courtesy Radio Shack, Division of Tandy Corp.).

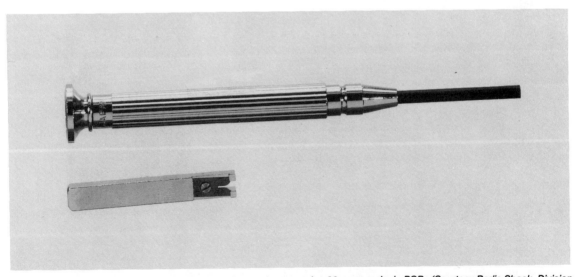

Fig. 5-14. This tool is used for cutting, wire-wrapping, and unwrapping 30-gauge wire in PCBs (Courtesy Radio Shack, Division of Tandy Corp.).

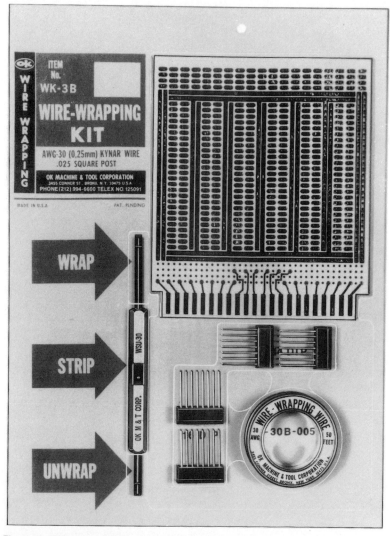

Fig. 5-15. This wire wrapping kit is complete with wrapping tool, wire, hobby board, and DIP sockets (Courtesy OK Machine and Tool Corp.).

removes 24- to 40-pin ICs, and the inserter inserts 14- to 16-pin DIPs with narrow profile. There is a pin straightner that aligns bent pins built into each tool.

SOLDERING

The art of soldering is at least 2000 years old, when the Romans made lead pipe by forming narrow strips of sheet lead into tubing and

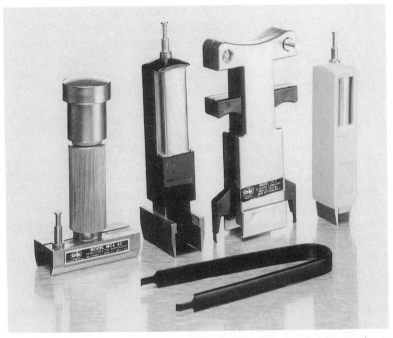

Fig. 5-16. An assortment of IC inserter and extractor tools that also straighten out bent IC pins (Courtesy OK Machine and Tool Corp.).

soldering the long seam. This soldered seam made for a watertight pipe. Later, the long-term preservation of food was made possible by the invention of the hermetically sealed tin can, which has been little changed for over 100 years. And of course, stained glass windows in medieval cathedrals were soldered together using lead. The durability of soldering is evident in some of these stained glass windows that have held up against stormy weather for over a thousand years.

Soldering is a quick, simple, inexpensive method of electrically joining two metals permanently. In any electrical circuitry that is mounted on a circuit board, the solder joint connects a component to the board both electrically and mechanically. The integrity of the solder joint is therefore critical to continued circuit operation.

Solder

A filler material is required to ensure a good electrical (low resistance) contact as well as some mechanical contact (rigidity) in soldering. This filler material—*solder*—melts at a lower temperature than the two metals to be joined together, for example, a copper terminal and a resistor lead. Heat is applied to the combination by means of a soldering iron, induction coil, blow torch, or hot bath. For electronics and electrical assembly such as printed circuit boards, the solder used is an alloy of tin and lead. Solder is usually identified by its percentage of tin and lead, such as 60/40, with tin as the first metal specified (the solder is 60 percent tin and 40 percent lead).

Solder Melting Temperature

The more tin in a solder, the lower the melting temperature. (See Table 5-1). Solder of 40/60 melts at 460 degrees Fahrenheit, and 50/50 solder melts at 420 degrees, etc. It is interesting to note that pure tin melts at 450 degrees, but lead melts at 620 degrees, a rather higher temperature. But when the two are mixed in the proportion of 63 parts of tin to 37 parts of lead by weight, a fusible alloy results (solder), which melts at a lower temperature than either metal—361 degrees. This latter composition has the lowest melting point of the tin-lead alloy.

Table 5-1. Tin-to-Lead Ratio Melting Temperatures in Solder.

Application	%Tin	%Lead	Melting Temp (degrees F)	Flux
General purpose work such as plumbing, gutters, car radiators, and other non-electrical repairs	40	60	460	Acid
General electrical work, appliances, house wiring, motors	40	60	460	Rosin Core
General non-electrical repairs, plumbing, radiators, etc.	50	50	420	Paste
Electronic repairs, TV, radio, PCB, kits	60	40	374	Rosin Core

The more tin in the solder, the more expensive the solder. In the United States, we tend to think of the tin can as being inexpensive. However, tin is expensive, and most tin cans are actually tin-plated steel cans. There are no tin mines in the United States, so it is imported from countries in Africa, Asia, and South America. Lead, however, is very inexpensive and mixes well with tin in all proportions to form an inexpensive solder.

Flux

There are three material things necessary for proper soldering:

➤ The proper flux
➤ Quality solder
➤ Sufficient heat

Some metals are easy to solder, such as tin and silver, because solder adheres well to both. Other metals, such as copper, iron, steel, monel, and nickel tend to oxidize easily when heated, so the solder does not readily adhere to them. In these cases, a chemical known as *flux* must be used while soldering to remove any oxide films that might have formed on the metals. When it comes to soldering, the skill of the solderer is of secondary importance to the use of the proper materials, and soldering flux is one of these important materials.

Most metals (except gold and a few others) oxidize, or tarnish, when exposed to the atmosphere. Copper turns green with copper oxide when it has been exposed to the elements for a long time. The thickness of the film or tarnish increases as time goes on, and with moisture or heat present, the oxidation process is speeded up greatly. Even if you can't see the tarnish, you will know it's there when it becomes difficult to get solder to adhere to such metals. The solder must *wet* and penetrate into the pores of the metal. The soldering flux allows this wetting to occur.

Flux does several things. First, it removes tarnish or metal oxide from the metal to be soldered. Second, the flux prevents further oxide from forming on the metal while it is being heated to soldering temperature. Third, it lowers the surface tension of the molten solder, enabling it to spread throughout the area that has been heated by the soldering iron.

It is important to remember that flux does not remove paint or dirt, only oxide films. Paint, dirt, and any other foreign material should first be removed by scraping, sandpaper, ultrasonic bath, or chemically.

There are many kinds of soldering fluxes—some good, some bad, and some very bad. One of the very bad ones for electronic or electrical construction is acid flux. The common fault of corrosive fluxes such as acid flux is that after you have finished the soldering operation, the flux residues attract moisture unless you remove them. In due time, the moisture sets up a galvanic action like that of a battery, and this causes the nearby metals to corrode. Using acid-core solder with a heavy iron is acceptable when doing *heavy* work such as soldering roof gutters and radiators, but for delicate electronic work, don't use it. Refer to Table 5-1 for a listing of melting temperatures and fluxes for different applications of solder.

Rosin—The Right Stuff

Rosin is a type of flux that is used to make up the core of some solders. Rosin works best on clean tin plates, clean copper, copper wire, pre-tinned wire, and solder-coated surfaces such as a metal chassis, however, metals that are difficult to solder are aluminum, iron, and cast iron. Rosin is desirable because the residue of rosin left from soldering is inert, non-corrosive, and it acts like an insulator. A rosin-core soldered joint ordinarily does not have to be cleaned unless an excessive amount of rosin remains. The rosin residue does not attract dust or moisture, and it will not corrode delicate electronic wiring.

DESOLDERING

During troubleshooting of electronic components boards, it is often necessary to remove a soldered component such as a transistor, IC, resistor, or capacitor. One of the first things you will discover is that putting them in is a lot easier than taking them out! However, there are several things that can make your job easier when removing a soldered component.

Heat and Pull

To remove the component, say a resistor, heat one end of the resistor with your iron until the solder melts. When the solder has melted or softened, use a pair of needlenose pliers to pull that end of the resistor loose. Then do the same desoldering process for the other end of the resistor. To allow the heat to flow more readily from the iron to the joint, put a bit of solder on the iron. When pulling the ends loose, be careful you don't splatter solder on yourself or other parts of the circuit.

Heat and Tap

In order to remove excess solder from a component that you might want to use again, heat the component leads and tap it lightly on a piece of wood on your bench and the solder will drop free. It is a good idea to clean off and open all connection holes in sockets, binding posts, etc. before reinserting and soldering components to them.

Solder Sucker

The solder sucker does just what its name implies—it sucks up solder from a joint that has been preheated and melted by your iron. The solder sucker is usually spring-loaded and causes a quick vacuum when released that sucks up the melted solder. It can be used as often as necessary to remove solder from a joint. The solder sucker is also available in the form of a rubber bulb that produces the suction.

IC Desoldering Tip

An iron-clad, pre-tinned in-line IC desoldering tool (Fig. 5-17) is available for melting solder on ICs. This tip is screwed into a standard ⅛- or ¼-inch thread in the soldering iron and encompasses both sides of the IC to heat all the leads. After the solder has melted, the IC can easily be pulled out.

Desoldering Braid

Desoldering braid is an interesting and scientific means of removing heated solder from a soldered joint. This braid is especially made for removing solder by means of *capillary* action. The braid can be placed across the pins of an IC chip and heated with a regular soldering iron. As the solder is heated, it melts and readily flows into

Fig. 5-17. IC desoldering tip used to desolder 14- or 16-pin DIP chips.

Fig. 5-18. Rolls of desoldering braid or wick are used to suck up melted solder from electronic components or boards (Courtesy Plato Products, Inc.).

the braid. You then simply remove the braid and all the solder from the connections comes with it. Once used, that part of the braid with cold solder in it is simply discarded by cutting it off.

An assortment of desoldering braids (or wicks) is shown in Fig. 5-18. These wicks are of several different sizes to work with the different types of electronic boards or components. The smallest braid is used

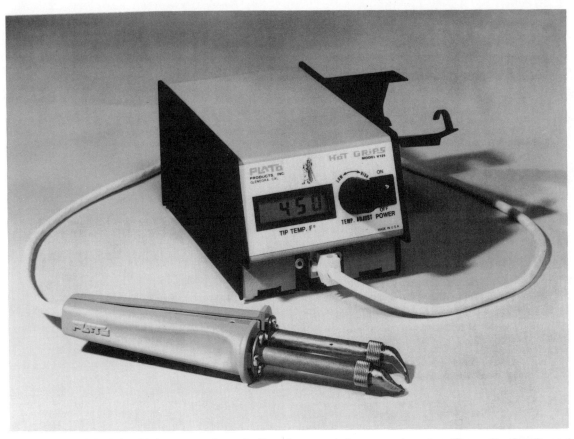

Fig. 5-19. "Hot Grips" desoldering tweezer is used with surface-mounted components such as ICs, small outline ICs, and small outline transistors (Courtesy of Plato Products, Inc.).

for fine electronic repair work, while the larger braid is used for production line repair or maintenance.

Hot Tweezer

Figure 5-19 shows a desoldering tool that can be used to quickly and safely remove surface-mounted components. Interchangeable tool tips are available to desolder IC chips, small outline ICs, and small outline transistors. The tip temperature is shown on a digital display and can be easily regulated. The "Hot Grips" is a hot tweezer tool made by Plato Products, Inc., 2120 East Allen Avenue, Glendora, CA 91740.

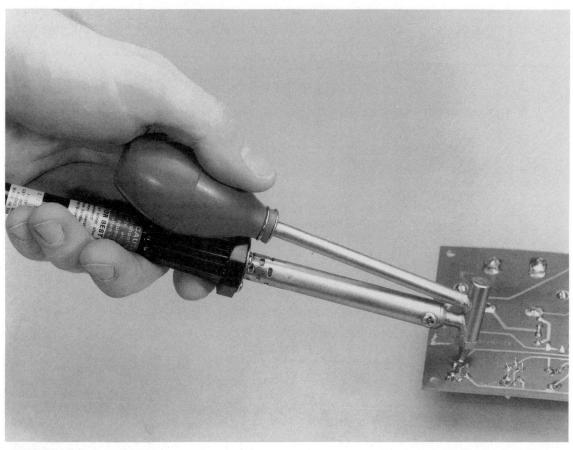

Fig. 5-20. A "one-hand" desoldering iron has a vacuum bulb to suck up melted solder (Courtesy Radio Shack, Division Tandy Corp.).

Desoldering Iron

To make it easy to remove and replace electronic components from circuit boards, Fig. 5-20 shows a "one-hand" desoldering iron. This iron features a vacuum bulb that sucks up the solder after it has been heated by the desoldering assembly and tip. The tip is heated by a 45-watt heater with a 5-foot ac cord. The whole assembly is 8 inches long. Replacement tips are also available.

Desoldering Station

A self-contained desoldering station with a high vacuum is shown in Fig. 5-21. The desolder tool has a quick-cleaning solder collector,

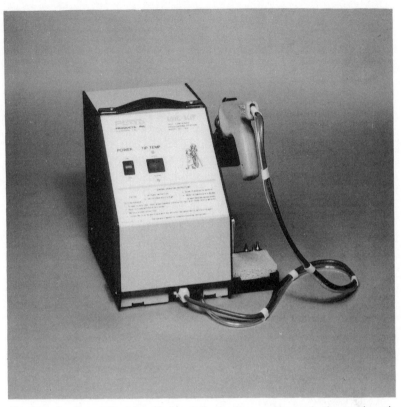

Fig. 5-21. A self-contained desoldering station is often used for production repair work (Courtesy Plato Products, Inc.).

a non-clogging tip, and conduit that can handle the toughest desoldering job. These types of stations are frequently used for production work.

TECHNIQUES

Even though present-day common soldering methods and techniques are very reliable, the great number of solder junctions on a board makes soldering the weakest link in the reliability of most electronic circuits. It is estimated by the circuit manufacturing industry that there are 200 million printed circuit boards assembled each year. The average board has about one thousand solder joints each, so there are an estimated total of 200 billion joints soldered each year. The Institute for Interconnecting and Packaging Electronic Circuits reports that 26 percent of the cost of assembling printed circuit boards is spent for repair and rework. Another 30 percent covers testing the board after rework. When the price of field repairs, test, and re-test on in-circuit equipment is included, defective solder joints can cost between $1.00 and $2.00 each. This amounts to an overall cost of $400 million lost each year due to poor solder joints and connections. Good soldering techniques are indeed important!

With patience and practice, you can become good at soldering anything, contributing to the operational reliability of your board. Keep in mind the following techniques when soldering.

➤ Avoid moving the parts to be soldered so the solder has a chance to cool after flowing throughout the area.
➤ You can use a vice or pair of pliers to hold a component while soldering, but don't let these larger metal tools absorb too much heat. You want the heat to flow to the metals to be bound—not used as a heat sink. When working on a metal bench top, use a piece of wood for holding the circuit board or component to avoid loss of heat.
➤ Hold the flat or broad side of the iron tip to the junction to quickly heat the area, and then apply solder to the other side of the junction with your other hand. The entire junction should heat to the point to where the solder melts without it ever even touching the iron. Solder should flow over the whole heated area. Be sure to hold the work at a level attitude so the melted solder doesn't run.
➤ Let the new junction remain steady for a few seconds until the solder has had a chance to solidify and set.

Avoid circulating too much air around the joint before it has a chance to solidify, because this can cause a "cold" solder joint. Sometimes these cold solder joints have a cloudy appearance, but frequently they have a normal appearance, making them difficult to locate when a problem arises. Tiny fractures can form in the joint that weaken it. Never use water to speed up the cooling process. There is no danger of corrosion forming from the rosin residue, though you might clean the soldered joint with a small brush or knife edge and then wipe the area with a clean cloth.

Tinning

Soldering iron tips must be smooth, and when soldering, you should "tin" (coat with solder) the tip so that heat can be easily transferred from the tip to the work to be soldered. To tin the tip of the iron, it must be heated long enough to the point where it changes color. Then touch rosin-core solder to it. The solder should flow easily over the tip so it has a silvery look.

Detinning

Detinning means the tip has lost its ability to hold solder properly, and the tip usually has to be replaced. Some of the common causes of detinning are:

➤ Failure to wipe and retin the tip properly.
➤ Lack of flux in the soldering operation, usually in touch-up and rework.
➤ Use of little solder, meaning the tip cannot be quickly flooded with solder.
➤ High tip temperature when idling, usually above 850 degrees Fahrenheit.
➤ Impure solder. Always use the best.
➤ Allowing foreign materials to get on the tip and burning, such as plastics, epoxies, and cements.
➤ "Wicking," which removes solder from the tip as well as the soldered joint.

Soldering Printed Wiring Boards

When soldering printed wiring boards, it is important to use the proper techniques and not apply too much heat or pressure to the work. If you do, you can cause a conductor pad to lift off the board, high temperature might burn it, or you could even loosen a terminal from the board. To avoid applying too much heat, use a 25-watt iron or smaller. To get to narrow spaces for soldering, use a small chisel tip (flat) or conical point. Apply just enough heat to allow the solder to flow to the connection. Remove the iron quickly without disturbing the joint so the solder can solidify and give you a neat connection. Note the technique in Fig. 5-22 for attaching a wire to an eyelet in a PCB.

Because soldering to PCBs is so important and is a crucial operation for good circuit reliability, the following precautions should be observed.

➤ Apply minimum heat so as to not burn the board or wire insulation.
➤ Do not disturb other components on the board with the hot iron.
➤ Do not let components move during soldering.
➤ Remove the iron as soon as the solder melts.
➤ When removing the iron, slide it off the connection. Do not pull up, because the copper foil could lift off the board.

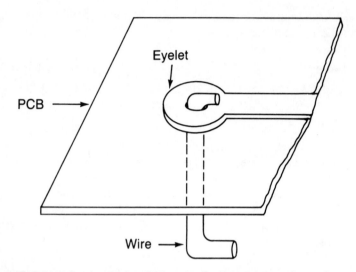

Fig. 5-22. Soldered connection to a PCB eyelet. Bend the wire over for maximum solder contact with the foil.

Preparing Wires and Terminals for Soldering

Wiring used in electronic circuit boards and assemblies is of two kinds—single-strand and stranded. The stranded wire is more flexible because it is made of six or seven small copper wires that have the same equivalent size as the single-strand wire. A number 22 stranded wire is the same size as a number 22 wire. An insulating material of plastic or fiber covers the wire that is stripped off before soldering the wire to a terminal. The amount of insulation to be stripped depends on the job to be done, but a usual length is about ½ inch.

Before soldering stranded wire to a terminal, you should tin the newly exposed copper. Hold the wire tip to the soldering iron and after a short while, touch the solder to the heated wire. The tinning process holds the stranded wires together and makes it very easy to solder to the terminal when it is heated. Always heat the wire sufficiently before applying solder to it. If heated properly, the solder should flow readily and you will have a neat and properly soldered connection. Try to avoid the temptation of putting solder on the iron rather than on the connection. Apply solder to the iron tip only when you are tinning it and not as a means of soldering. When the terminal or wire has been heated properly, the solder will flow freely. Then quickly remove the solder and the iron so the terminal and solder don't overheat. In addition, overheating the junction can melt the insulation of the wire or possibly cause solder to flow under the insulation.

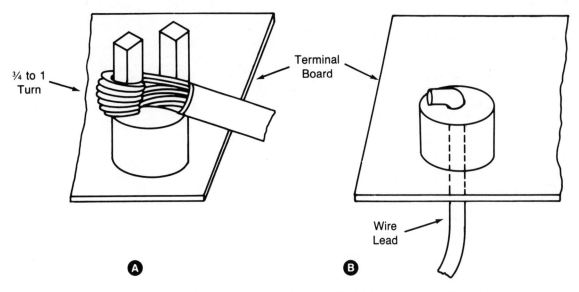

Fig. 5-23. Wire wraps around terminals prior to soldering around a prong (A) and through a terminal board (B).

Soldering Wraps and Splices

In printed circuit or electronic chassis wiring, be sure to first have a good mechanical wrap or splice around the wire or terminal. The mechanical integrity (ruggedness) depends on how you have connected the wire to the terminal. You should not depend on the solder applied to hold the connection together. The solder is there to provide a good electrical connection only.

Wire Wraps. The wire to be connected to a terminal should be stripped back about ½ inch and then wrapped at least ¾ turn as shown in Fig. 5-23A. Remember to pre-tin the wire before wrapping it around the terminal. The solder will then flow very easily when soldered.

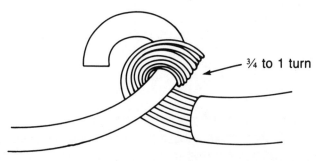

Fig. 5-24. Wire wrap around hook terminal. Note gap between insulation and terminal.

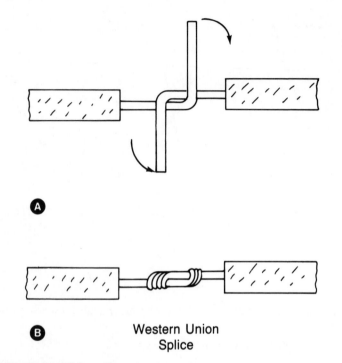

A

B   Western Union
                    Splice

Fig. 5-25. Western Union splice.

Wire Splices. When splicing two wires together, it is again best to assure yourself that you have a good mechanical connection before applying solder. Figure 5-24 shows how two wires are connected together using a hook splice. Make a small hook in each wire, hook them together, and then squeeze them with a pair of pliers. You can then solder the connection.

A splice that provides excellent mechanical connection between two wires is the Western Union splice. Figure 5-25A shows where the two wires are crossed at right angles. The wires are then wound around each other as shown in Fig. 5-25B which assures good mechanical rigidity. If the wires are pretinned, the joint will solder very easily when heat and solder are applied. Remove the iron as soon as the solder begins to flow, and wait a few seconds for the solder to cool and solidify before moving the connection.

Repairing Damaged or Missing Tracks on PCBs

If the copper foil comes loose from the PCB due to overheating, (see Fig. 5-26) there are several repair kits available to come to your

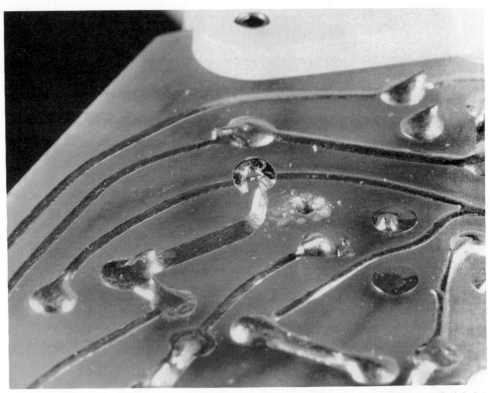

Fig. 5-26. The pad and trace have lifted off the board, probably because too much heat was applied during soldering or rework of the board in that area (Courtesy PACE, Inc.).

rescue to affect a speedy repair. One such repair kit is from Pace, Incorporated, known as the CIR-KIT. This kit provides an efficient, low-cost way to repair and/or replace lifted, damaged, or missing lands, and plated-thru holes, conductors, and edge connectors on printed circuit assemblies. CIR-KITS can help reduce downtime and eliminate the sometimes costly practice of discarding PCBs that can be easily and quickly repaired.

The kits are easy to use, because everything you need for repair on assembled PCBs is included. The tracks are plated to military specifications, and they come in sheets of various pad diameters and track widths for any type of repair. Included are over 30 sizes of eyelets, an abrasive stick, setting tool for cold-setting eyelets, and detailed, simple-to-follow instructions.

Figure 5-27 shows a "donut pad" and conductor trace that have been placed on the PCB and are about to be soldered to the board, and Fig. 5-28 shows that the completely repaired portion has been

Fig. 5-27. The raised trace and donut pad are cut free and a replacement pad is inserted (Courtesy PACE, Inc.).

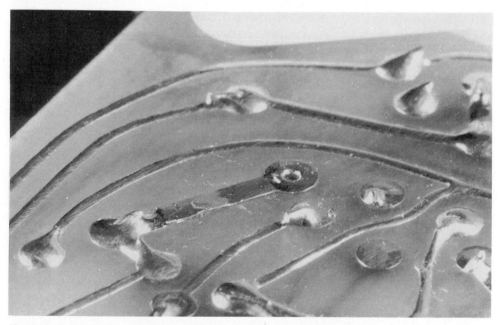

Fig. 5-28. The pad and copper foil are soldered in place (Courtesy PACE, Inc.).

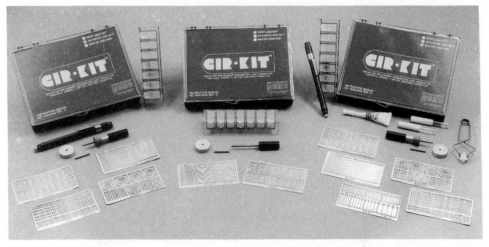

Fig. 5-29. The CIR-KITS shown allow repair of printed circuit boards where foil has lifted off the board (Courtesy PACE, Inc.).

replaced and looks as good as the rest of the board. The component lead can now be inserted and soldered as easily as before.

The Pace CIR-KITS are available in three different models, including the Basic, Advanced, and Master CIR-KIT. These kits suit various types and levels of repair in the factory or for field use. The three CIR-KITs are shown in Fig. 5-29 with working materials provided for repairing lifted copper foil. (CIR-KITs are available from PACE, INC., 9893 Brewer's Court, Laurel, MD 20707. Telephone is (301) 490-9860.)

CHAPTER 6

Types of Wiring and Circuit Boards

There are some important considerations to adhere to when designing, building, and working with circuit boards. The most important of these is *safety*. Avoid sharp corners and edges. Remove any rough edges you might have on an insulated board or metal chassis with a file or sandpaper before you start to work on it. Avoid awkward shapes that might cause you to mishandle and drop the board. You should be able to easily reach all parts and components mounted for any necessary troubleshooting and repair.

Mount parts so there is no danger of shock from exposed high voltage. Also, in your design, think of any damage that might be caused by high voltage if a part failed. Warning signs such as "High Voltage" or "Warm When Operating" should be placed near those components to be avoided.

When it is not necessary, don't pack your components too closely together on a board or chassis. They are more difficult to remove when a repair is necessary, and close packing might increase your heat dissipation problems. Keep parts spread out if at all possible when circuit operating speeds are not a crucial design consideration.

Above all, consider *neatness*. Lay out the parts so your circuit looks neat when wired or connected. A neat circuit arrangement not only looks more professional but it is more likely to work more efficiently and is easier to troubleshoot.

WIRE DESIGNATIONS

Wires are used to interconnect point-to-point circuit boards or to connect one board to another through a wiring harness. To identify the functions of the various wires, color codes are used on the insulation of the wires.

The color coding used in the industry is identical to that used to describe the values of resistors and capacitors, as described in Chapter 3. If only 10 wires are involved in a circuit or interconnecting harness, only one list of colors is required—from black to white (0 through 9). An orange wire is identified, or designated, 3, a blue wire 5. Table 6-1 shows the numerical assignment ascribed to each color combination so that a number of wires above 10 can be easily identified. In this manner, hundreds of wires can be separately identified in a large cable, such as is used in the telephone industry. For example, a red wire with a blue stripe (tracer) is designated as 25. In order to designate a red wire in the two-number scheme, a 22 would be used (red, red). To designate an orange wire in the three number scheme, 333 is used (orange, orange, orange) to indicate an orange wire with two orange tracers.

WIRE SIZES

Wire "size" is the cross-sectional area of the wire that determines how much current the wire can safely carry without overheating. For

Table 6-1. Color Code for Wire Designation.

Color	Solid	Stripe	Stripe
Black	0	0	0
Brown	1	1	1
Red	2	2	2
Orange	3	3	3
Yellow	4	4	4
Green	5	5	5
Blue	6	6	6
Violet	7	7	7
Gray	8	8	8
White	9	9	9

Table 6-2. Copper Wire.

AWG#	Area (Mils)	Ohms/1000 Feet	Max Current (amps)
0000	211750	.04900	302.6
000	167721	.06179	240.0
00	133163	.07791	190.3
0	105600	.09825	150.9
1	83742	.1239	119.7
2	66409	.1562	94.9
3	52663	.1970	75.3
4	42762	.2484	59.7
5	33118	.3133	47.3
6	26263	.3951	37.5
7	20827	.4982	29.8
8	16516	.6282	23.6
9	13097	.7922	18.7
10	10386	.9989	14.8
11	8237	1.26	11.8
12	6532	1.59	9.33
13	5180	2.00	7.40
14	4108	2.53	5.87
15	3257	3.19	4.65
16	2583	4.02	3.69
17	2048	5.06	2.93
18	1624	6.39	2.32
19	1288	8.05	1.84
20	1022	10.2	1.46
21	810	12.8	1.16
22	642	16.1	0.918
23	509	20.4	0.728
24	404	25.7	0.577
25	320	32.4	0.458
26	254	40.8	0.363
27	201	51.5	0.288
28	160	64.9	0.228
29	127	81.9	0.181
30	100	103.3	0.144
31	79.7	130	0.114
32	63.2	164	0.090
33	50.1	207	0.072
34	39.7	261	0.057
35	31.5	329	0.045
36	25.0	425	0.036
37	19.8	524	0.028
38	15.7	660	0.022
39	12.5	833	0.018
40	9.88	1050	0.014
41	7.84	1324	0.011
42	6.21	1669	0.0089
43	4.93	2105	0.0070
44	3.91	2655	0.0056
45	3.10	3348	0.0044

most electronic circuit wiring, small wires can be used. The smaller the wire, the higher the number. Table 6-2 shows the various copper-wire size numbers, cross sectional area, resistance, and how much current they can carry. The wire can be solid copper or stranded, but they have the same number. (In general, solid wire is more difficult to bend, so it is usually preferable to use stranded wire.)

POINT-TO-POINT WIRING

In electronic circuit assembly, there are two broad categories of wiring: (1) point-to-point, either by hand or wire-wrap and (2) printed or etched wiring. As discussed earlier, in point-to-point wiring, individual lengths of insulated wire are connected between components or between components and the chassis. This means the components are first inserted into the board and then the wires are connected from one point on the board or chassis to another. If the component leads are long enough, they can be used as the interconnecting wires. When the wire is in place and a good mechanical connection made, you can then solder to complete the electrical connection. Do this with each connection as you work your way through the circuit diagram.

When you wire a circuit by hand, be sure that all wires are parallel and square to the board to present a neat appearance. Be sure to isolate the wires from moving parts, such as a variable tuning capacitor or potentiometer. If a wire passes through a metal chassis hole, be sure to provide isolation from vibration and ensure proper insulation by using a grommet.

Before explaining printed wiring, a discussion of alternate mounting boards—breadboarding and perforated boards—is covered next.

BREADBOARDING

Before committing a circuit design to a PCB, some designers lay their circuits out on a breadboard to make temporary connections. The term "breadboard" is still used today even though it is over 50 years old. It stems from the early days of radio building when a breadboard was borrowed from the kitchen to mount radio components onto temporarily. After the circuit operated properly, it would then be assembled into a metal chassis.

As a fast way to design a circuit, the solderless modular breadboard accepts a wide variety of devices and wires by simply plugging and unplugging them. Shown in Fig. 6-1 is a modular IC breadboard socket that has silver-nickel contacts that accept wire sizes from #30 to as large as #22-gauge solid wire. The board has two bus

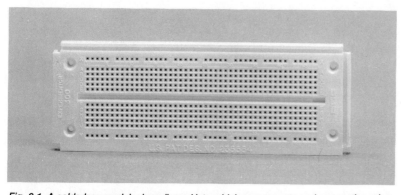

Fig. 6-1. A solderless modular breadboard into which components and connecting wires are inserted to complete a circuit (Courtesy Radio Shack, Division of Tandy Corp.).

strips and 550 indexed connection points. The units snap together horizontally or vertically for larger or more complex designs.

A universal solderless breadboard is shown in Fig. 6-2. This 2¼ × 6½-inch breadboard is mounted on a 7 × 4-inch steel base with rubber feet so the unit stays put when you insert components. The board accepts DIPs, discrete components, and wire up to #22-gauge. There are a total of 640 plug-in tie points and three binding posts for external power connections (two different voltages and ground).

Solderless breadboarding units are very useful in the electronics industry and home experimenter's array of equipment. Figure 6-3 shows a breadboarding unit that combines terminal and distribution building blocks in a single unit. The unit has the plug-in ease of the 0.1 × 0.1-inch solderless tie-point matrix combined with the convenience provided by separate distribution buses for power, ground, and signal lines.

A high-performance low-inductance breadboarding system is shown in Fig. 6-4. This board has two levels of printed circuits, three front-mounted binding posts and three distribution buses. It has a total of 3208 plug-in tie points. The unit is engineered for TTL (transistor-transistor-logic) and CMOS (complementary metal-oxide semiconductor) circuitry.

Jumper wires are used to connect the various components together and are easily inserted and removed as circuit components are changed. An assortment of these jumper wires is shown in Fig. 6-5.

Other plug-in breadboards are shown in Fig. 6-6, where the smallest board can mount six 14-pin DIPs and the largest can mount up to 54 14-pin DIPs. The large board measures approximately 15 × 10 inches.

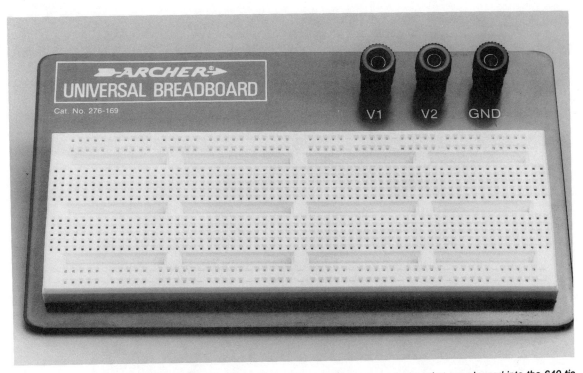

Fig. 6-2. This universal solderless breadboard can accept various components and wires that are plugged into the 640 tie points (Courtesy Radio Shack, Division of Tandy Corp.).

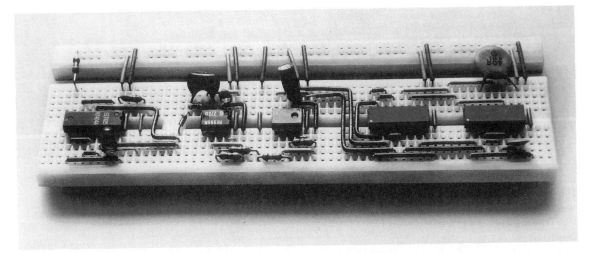

Fig. 6-3. A solderless plug-in breadboard allows for easy design of a circuit before turning it into a PCB (Courtesy AP Products, Inc.).

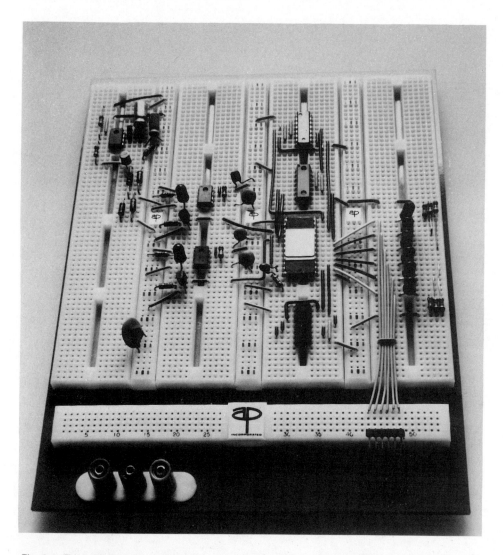

Fig. 6-4. This breadboard has 3208 plug-in tie points and two voltage distribution systems (Courtesy AP Products, Inc.).

The plug-in breadboards discussed above can be connected together by DIP jumpers shown in Fig. 6-7. These flat ribbon cable assemblies have molded-on DIP connectors used for jumpering within a PCB, interconnecting between PCBs, backplanes, and motherboards, and interfacing input and output signals. The cables are available in 14, 16, 24, and 40 contact sizes. Standard lengths are 6-, 12-, 18-, 24-, and 36-inch. Installed IC chips can then be tested

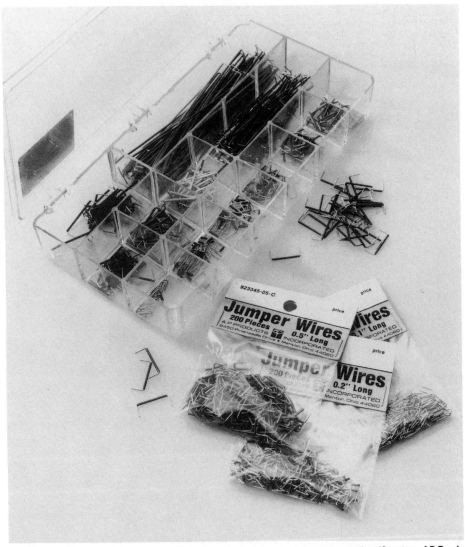

Fig. 6-5. Plug-in jumper wires are used to connect components and points together (Courtesy AP Products, Inc.).

by using a clamp-on test clip as shown in Fig. 6-8. The leads are then run to an oscilloscope or test meter.

PERFORATED BOARDS

When designing circuit boards, it is handy to have a perforated board on which you can mount components to try different

93

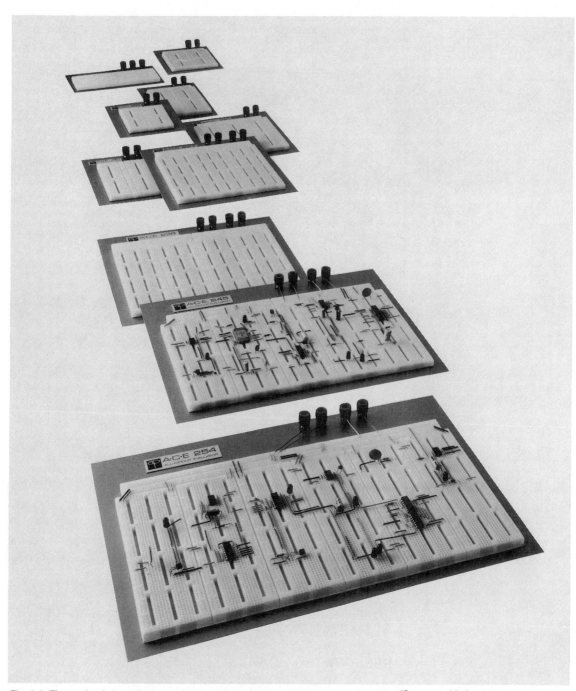

Fig. 6-6. These plug-in breadboards will accept from six 14-pin DIPs to 54 14-pin DIPs (Courtesy AP Products, Inc.).

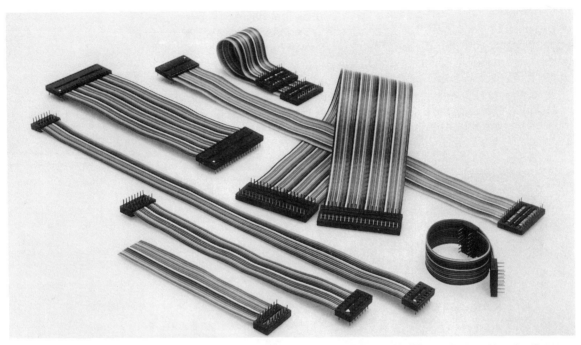

Fig. 6-7. Plug-in tie-point boards are interconnected by DIP jumper ribbon cables of different sizes and lengths (Courtesy AP Products, Inc.).

Fig. 6-8. A clamp-on test clip is used to test ICs while they are mounted in the PC board. Leads can be connected to a scope or meter (Courtesy AP Products, Inc.).

arrangements. Figure 6-9 shows a "perf board" that has a number of pre-drilled holes in which you can insert component leads while experimenting with your design. The board shown has pre-punched holes 0.042 inches in diameter in a standard 0.1 × 0.1-inch grid. The board has columns and rows that are indexed to identify component pinouts and has 25 vertical rows of 30 holes each. The hole spacing is designed to accept the 0.1-inch pin spacing of the standard IC DIP (dual in-line package). The chips can be placed horizontally or vertically anyplace on the board.

Fig. 6-9. A perf board for mounting electronic components (Courtesy Radio Shack, Division of Tandy Corp.).

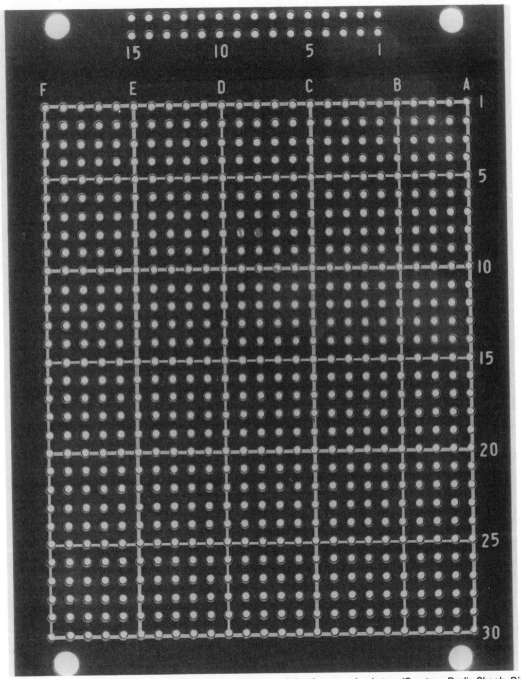

Fig. 6-11. A dual mini general-purpose board is shown with a break line for snapping in two (Courtesy Radio Shack, Division of Tandy Corp.).

Fig. 6-11. A dual mini general-purpose board is shown with a break line for snapping in two (Courtesy Radio Shack, Division of Tandy Corp.).

Figure 6-10 shows another perfboard that is lettered across the top from A to F and numbered down the side from 1 to 30. This helps to keep track of which components go where, inserting a resistor in B5 and C5, for example.

A dual mini general-purpose perfboard is shown in Fig. 6-11. Each side of the board has 213 holes and measures 1¾ × ⅝ inches. The board can be snapped in half for smaller circuit requirements.

PRINTED WIRING

Printed wiring boards are often incorrectly referred to as *printed circuit boards*, because in printed wiring, only the *conductors* are printed (not the components). After that, components are attached to the board and are soldered in place. However, the two terms are often used interchangeably. A *printed circuit card* (PC card) is a removable circuit board that "plugs" into a device, such as a personal computer.

The printed wiring board is made of a non-conducting laminated material to which a thin sheet of copper has been bonded on one or both sides. The foil is usually quite thin, from 0.0007 inches to as thick

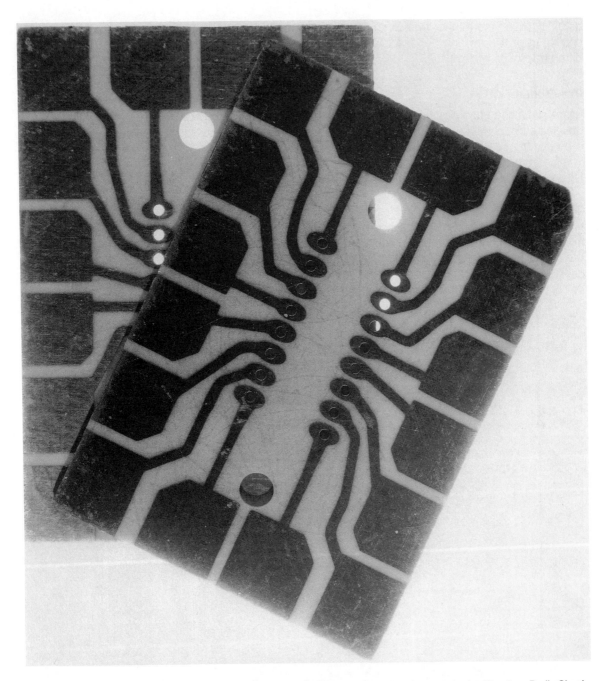

Fig. 6-12. A pre-etched PWB for a 16-pin DIP IC chip. Connections are soldered to the copper tabs (Courtesy Radio Shack, Division of Tandy Corp.).

as 0.010 inch. The thinner the foil, the less current it can carry. The circuit design is drawn on the foil and etched onto the board by several different processes.

Pre-Etched PCBs

A pre-etched board is a general-purpose board on which you can develop a circuit without designing a custom board. Figure 6-12 shows a board that can be used by experimenters to mount a 16-pin DIP. The board can be sawed in half so that it can be used with an 8-pin DIP, such as the popular 555 timer IC.

A dual IC board is shown in Fig. 6-13 that can accept two 6- to 20-pin DIPs. The board also provides four connecting points for each IC pin. An experimenter's PCB is shown in Fig. 6-14. The layout matches the modular breadboard socket shown in Fig. 6-1. It has 556 pre-drilled, indexed connection points and measures $2\frac{1}{16}$ × $5\frac{7}{16}$ inches. A multi-purpose plug-in board is shown in Fig. 6-15. It measures

Fig. 6-13. A pre-drilled PC board for two 6 to 20-pin DIP IC chips (Courtesy Radio Shack, Division of Tandy Corp.).

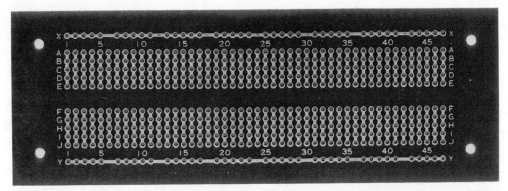

Fig. 6-14. This PC board is used to permanently wire up a completely circuit that was tested on a breadboard (Courtesy Radio Shack, Division of Tandy Corp.).

4¼ × 4 inches with ¹⁄₁₆-inch grids and has three buses. The board plugs directly into a 44-position card-edge socket.

Custom-Etched PCBs

A custom-etched PCB is one that is specially etched for a particular circuit design. After all the bugs or problems in a circuit have been worked out on an experimenter's board, lay the circuit out on a custom PCB that you can etch or print yourself. Greater details follow in Chapter 7, Circuit Design and Board Layout.

COMPONENT MOUNTING

Components are mounted on one side of the board, opposite the foil conductor traces. See Fig. 6-16. Recall that in point-to-point wiring, components are mounted first and then wires are attached to complete the connections, but in printed wiring, the conductors are etched (printed) onto the board first and then components are mounted, or *stuffed*, and soldered to the wiring. Though the two methods are different, the same general rules apply for mounting components. (See Fig. 6-17.)

1. Mount polarized components such as capacitors and diodes in the proper directions.
2. Arrange color-coded components such as resistors, capacitors, and semiconductors so that codes can be easily read from left to right or from top to bottom.
3. Arrange all components and parts in parallel lines if possible to present a neat appearance as well as ease of maintenance.

Fig. 6-15. This multi-purpose plug-in board has three buses (Courtesy Radio Shack, Division of Tandy Corp.).

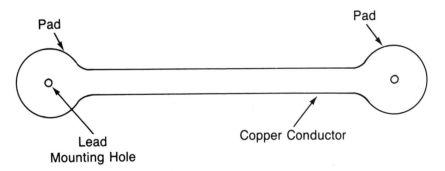

Fig. 6-16. *Copper foil conductor and soldering pad on PCB.*

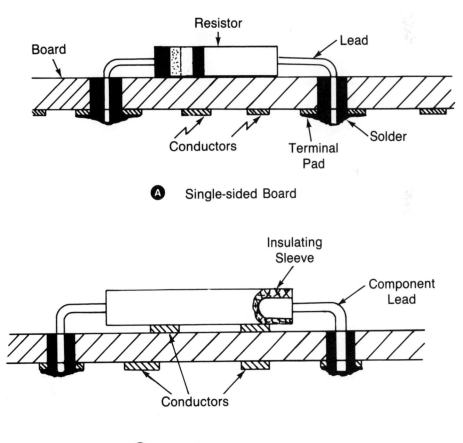

Fig. 6-17. *Component mounted on single-sided board (A) and double-sided board (B). Note insulating sleeve around component in (B).*

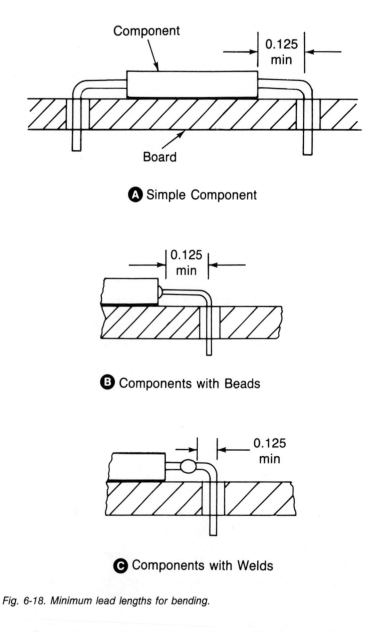

Fig. 6-18. Minimum lead lengths for bending.

4. Center the components between mounting holes unless space restrictions dictate otherwise.
5. Always make a good mechanical connection. This assures a good electrical connection when soldering or using a wire wrap tool.
6. Cut component leads to proper length after soldering.

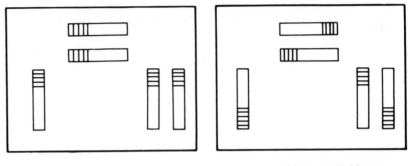

Acceptable   Unacceptable

Fig. 6-19. Arrange resistors parallel to each other and for ease of readability.

In Fig. 6-17, the component is mounted very close to the single-sided board. An insulating sleeve is placed over the component when mounted on a double-sided board so it doesn't produce a short across the foil. However, if the component is a resistor and is rated in the area of 1 watt of power, it is best to raise it about ⅟₃₂ of an inch from the board. Component leads should have a minimum horizontal length from the body to the pad hole of 0.125 of an inch, as shown in Fig. 6-18. Run the leads parallel to the board until they are to be bent to go straight down through the hole.

Arrange resistors and capacitors neatly on the PWB, placing them parallel to each other horizontally or vertically as shown in Fig. 6-19. Remember to place the resistors so their values can be read from left to right or top to bottom. Figure 6-19A shows an acceptable arrangement while Fig. 6-19B is not acceptable.

CHAPTER 7

Circuit Design
and Board Layout

How does one go about taking a schematic diagram for a circuit and laying it out on a board? This chapter explains some of the methods to do this.

PRINTED CIRCUIT BOARD LAYOUT

Parts and components should be laid out on a printed circuit board in a neat and orderly manner. To make efficient use of the circuit board, lay the components out in a manner as shown in Fig. 7-1. This is especially important if the board has many components mounted on it. Note from Fig. 7-1B how much more compact the layout is compared to Fig. 7-1A and the amount of board space saved by this approach. (In the arrangement shown, assume that power dissipation of resistors or other active components is not a factor, which allows tighter component packing.)

DISCRETE COMPONENT LAYOUT

When laying out discrete components on a PCB, the following items should be considered as they will simplify the design process.

➻ Work around or build from an important component such as a transistor or IC chip. Make this the center of your circuit.

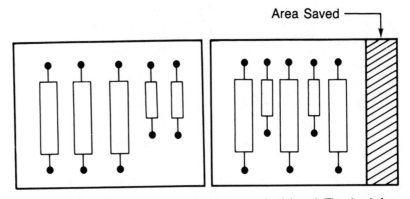

Fig. 7-1. Efficient parts placement can save room on circuit board. (The view is from the component side of the board.)

➥ Consider the shape of any grouped components, such as in a square or rectangular shape.

➥ Try to work component layout on the board the same as the schematic diagram—input on the left (or top) and output on the right (or bottom).

➥ Choose a board size large enough to easily accommodate all components.

➥ The case (size) of the component should be used to determine the lead spacing on the PCB.

➥ The lead diameter must be determined so a pad size can be selected.

➥ Determine orientation of the pad when looking at a top view of the components, such as a transistor's emitter, base, and collector.

➥ Determine polarity of all polarized components such as diodes and LEDs (identify the anode and cathode end), electrolytic capacitors, and transistors (identify the emitter, base and collector).

TRANSFERRING SCHEMATIC TO BOARD

Looking at a schematic diagram, it looks fairly simple. Figure 7-2 shows just a few lines and resistors connected to a transistor to make a simple audio amplifier. While the lines look simple, where do they really go? Look at the base circuit (components tied to the base) of the transistor in Fig. 7-2, for example. Figure 7-3 shows the base circuit redrawn. The base connects to R1, R2, and to C1. In effect, there are four components connected together at a common node.

As an overview, there are two components connected to the emitter of the transistor (R1 and C2), two connections to the collector

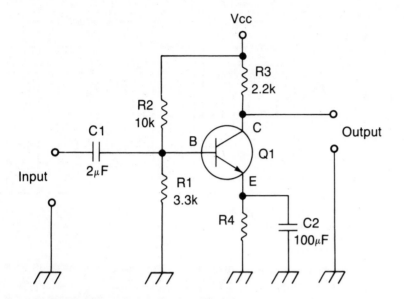

Fig. 7-2. Small-signal transistor circuit schematic.

(R3 and the output) and two components connected to V_{cc} (R2 and R3). Also, there are three leads connected to ground (R1, R4, and C2).

Transferring the components to a PCB yields the recommended layout in Fig. 7-4. Note that the layout is a rectangle or square, and Q1 is at the center because it is the major component in the circuit and connects to more components than any other component in the circuit. There is no one "perfect" way to lay out the components and wiring, but you should optimize the arrangement to make components and the input and output terminals easy to get to.

The layout is arranged so the signals move from left to right—input on the left, output on the right. The power supply is at the top

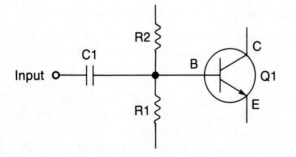

Fig. 7-3. Four components share a common node with the base of transistor Q1.

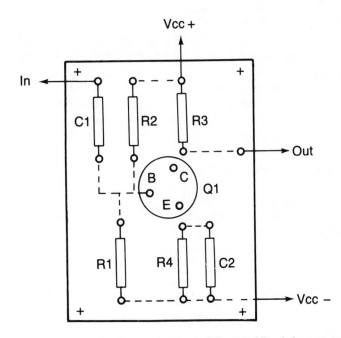

Fig. 7-4. Board layout of components for circuit of Fig. 7-2. (View is from component side.)

and ground is at the bottom. This is not absolutely essential, but this logic is a good place to start and will help in troubleshooting the circuit.

After you have finished a rough sketch of the proposed layout, check what you have against the following general guidelines:

➤ Arrange components on the PCB to achieve the optimum shape (rectangular (horizontal or vertical), square, round, etc.).

➤ Do any components require special structural or obstruction considerations (light sensitive components, etc.)?

➤ Are heatsinks required for certain components (transistors, high-power diodes, and switching devices such as triacs or SCRs?

➤ Minimize capacitive or magnetic coupling. Keep input and output transformers separated.

➤ High potential differences between certain conductors (signal input lines and supply voltages) make it advisable to maintain safe spacing between such conductors.

➤ Certain components should have short trace lines to minimize noise interference (input circuits, etc.)

➤ Components with metal outer cases should not touch each other as the case might be electrically "hot."

➳ Heat-sensitive and heat-radiating components should be located as far apart as practical. However, do not place all heat-generating components near each other as this could create a "hot spot" area.
➳ Watch for separate grounds that might be used on the same schematic, such as different power supply voltages.
➳ Heavy components should not be supported by their solder leads alone. Provide additional support through use of a bracket or clamp.
➳ Check to see that all components have the proper reference designators such as R1, C3, etc., and that all necessary components have been accounted for.

Finally, using a PC design template, draw the components to actual size. Use the maximum component size so all components fit onto the board, because a smaller component will fit in place of a larger one, but once laid out, a larger component cannot replace a smaller one.

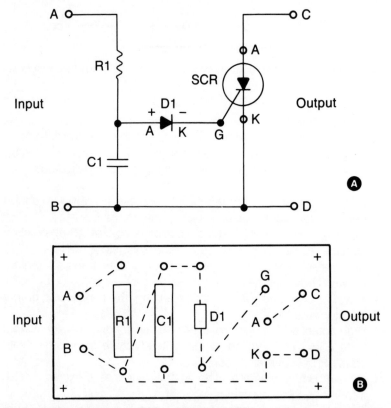

Fig. 7-5. Circuit diagram for an SCR switch (A) and PCB layout diagram of components and copper foil (B). (View is from component side.)

110

A REVIEW CIRCUIT

As another example, Fig. 7-5A shows a circuit diagram for a silicon-controlled rectifier that turns on a dc circuit (at the output) when a smaller dc voltage appears momentarily at its gate (input). This is a popular circuit for turning on a siren or piezo sounder.

In Fig. 7-5B, the schematic is depicted as a layout diagram on a PCB. Working from the input to the output of the circuit (left to right), note the following conditions and compare the schematic to the layout.

➤ R1 and C1 are across the input of the circuit (from A to B).
➤ The positive side (anode) of the diode, D1, is connected to the junction of R1 and C1.
➤ The negative side (cathode) of the diode, D1, is connected to the gate (G) of the SCR.
➤ The anode side (positive) of the SCR connects to terminal C of the output and the cathode (negative) of the SCR connects to terminal D of the output.
➤ Terminal B of the input connects to the negative side of capacitor C1 and to the negative (cathode) of the SCR, and the output terminal D.

By using the schematic as the master drawing, you have verified that the layout diagram agrees with the schematic.

THE PRINTED CIRCUIT BOARD PROCESS

Most PCBs begin as a layer of copper on an insulator such as fiberglass, as shown in Fig. 7-6. You must first place a protective coating on the copper where you desire the traces and islands to be. Then the unwanted copper can be removed by etching. Begin by making a *positive transparency* (or *positive*).

Producing the Positive Transparency

Producing a positive is a fairly easy task when using the wide range of drafting aids available from the various electronics suppliers. You don't really have to be an artist; you just have to be able to draw a reasonably straight line along a straightedge. The drafting aids help you to do the rest. For the beginning experimenter, it's best to use a single-sided board, because a double-sided board requires that all of the pads and holes line up (*reqqistration*) and can be tricky.

To begin the layout of the positive, tape a sheet of ¹⁄₁₀-inch grid paper to a drafting board. Next, place a piece of Mylar film over the top of the grid sheet and secure it with masking tape. Obtain an opaque,

Fig. 7-6. A variety of sizes of PCBs are available from local electronics stores (Courtesy Radio Shack, Division of Tandy Corp.).

precision drafting template that contains most of the popular components. Look for those drafting aids that have a translucent backing with pressure-sensitive "repeat" adhesive that enables you to lift and reposition the pattern. A number of very handy patterns are available, including pre-spaced donut patterns on $\frac{1}{10}$-inch centers.

Some representative patterns are shown in Fig. 7-7, which includes PC layout strips and circles, connector contact patterns, dual-in-line patterns, donut pads, 3-lead TO-5 patterns, 8-lead TO-5 patterns, and 10-lead TO-5 patterns. If you have an unusual component, chances are it can still be made with leads spaced on this (industry) spacing of $\frac{1}{10}$ of an inch. Pads spaced on 0.156-inch centers are also available. Hence, there are patterns for ICs, transistors, PC card-edge connectors, and a very wide variety of donut pad sizes for mounting resistors, capacitors, etc. See "Drilling the Board," later in this chapter.

When laying out the circuit on the board, place all of the component patterns and donut pads on your transparency with the help of a pair of tweezers. Next, position the interconnections between components or points using drafting adhesive tape. These drafting tapes are available in many widths, and it is best to use the largest practical width for several reasons. First, wider tapes are easier to apply to your transparency than the very narrow sizes. Secondly, they result in greater conductivity (carry more current) because of the wider conductor. Also, the wider conductor size results in fewer rejections of the finished board.

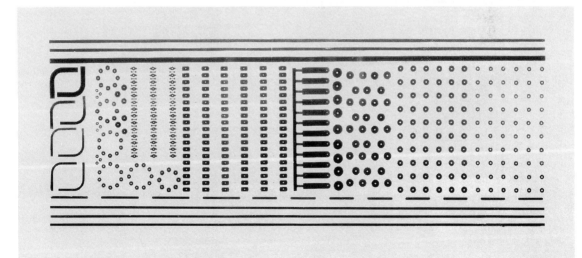

Fig. 7-7. PCB drafting patterns are available for a number of components. The aids can be repositioned if necessary (Courtesy Radio Shack, Division of Tandy Corp.).

Although the chemicals used for producing the PCB can produce extremely fine detail, the wider lines are much less subject to problems from dust particles, mishandling, improper exposure time, light source intensities, and the like.

When placing the tape on a donut, hold the roll with one hand and attach and hold down the cut end with the other. Then unroll the tape and run it to the next termination point and press down. Then, run a finger along the tape to assure an even surface. When you are ready to cut the tape over a pad, hold a sharp knife edge firmly in a fixed position in a straight line across the width of the tape. Then pull the tape up at an angle to the knife edge to assure a clean and easy cut. With long traces, run a finger along the tape as you roll it out to make sure it adheres to the Mylar. Soon you will have laid out your circuit on the Mylar positive and are ready for the next step. A reminder—most PCB products are based on photo sensitive chemicals, hence the component pads and tapes are totally opaque. The copper boards require this high opacity for best results.

Producing the PCB

After producing the transparency, you can now start the etching process. This section covers the following steps.

➤ Sensitize the PCB
➤ Expose the board and positive to ultraviolet light
➤ Develop the board
➤ Etch the board
➤ Drill the board for parts loading

Cleaning the Board. In order to have a good printed circuit board, the copper surface must first be thoroughly clean. Use fine steel wool, household cleanser, and tap water. It is very important to *not* use pre-soaped kitchen cleaning pads. Once you have cleaned the board, handle it only by its edges to keep from getting hand oils on the copper surface.

How do you tell when the board is really clean? A ''water break'' test will quickly tell you this. Water from a faucet should sheet evenly across the entire surface of the PCB with no tendency to pull away from the edges or form separate pools. When beads of water form on the surface (like those on a waxed car finish), it is a sure sign that the board is not clean. It may be necessary to clean the board once again. Once you have cleaned the board, it should be dried with a lint-free cloth.

Coating the Board. After you have cleaned the board, spray it with an aerosol positive-acting sensitizer. Lean the board against the wall so that it is standing nearly vertical. Spray the board from a distance of 8 to 10 inches with a back and forth horizontal motion and without releasing the spray valve. The spray should overshoot the ends of the board as you spray from bottom to top in a series of horizontal passes so that the board is covered evenly and doesn't run.

When the board is completely coated, lay it face up in a dark place such as a drawer or cupboard, to self-level the spray. Any excess spray at the edges of the board can be drained by touching the edge with a paper towel. Because of variations in user techniques, you might need to apply more than one coat. In this case, the board should be completely dried between each coat.

Drying the Board. The board can be drawer-dried overnight or force-dried in an ordinary oven set on low. The oven drying method is preferred and normally yields the most consistent end result.

For the oven-drying method, allow the board to rest horizontally for about 15 minutes in the dark before placing it in the oven. Pre-heat the oven on low to about 120 to 150 degrees Fahrenheit and be sure the oven is dark. A drying time of 20 to 25 minutes is recommended. After it has been heated, the board must be allowed to return to room temperature before handling. This takes about another 20 minutes.

Exposing the Board. Ultraviolet light is necessary for adequate exposure but varies, depending on the light source you use. Large professional equipment takes only 1 to 6 minutes, while a sunlamp at a distance of 12 inches takes 12 to 14 minutes. The sunlamp is the same as those commonly used for tanning skin.

To expose the PCB, place the Mylar transparency on top of the board and secure it under the light source. Be sure to hold the transparency firmly in place against the coated board using a sheet of glass or a contact film frame.

Developing the Board. Ready a tray of developer by using a mixture of three parts of water to one part of developer concentrate. GC Electronics No. 22-226 concentrate, available from local electronics suppliers, is one such developer. Be sure to use a glass or plastic tray and not a metal one. Fill the tray to a depth of about ½ to 1 inches with the developer and water solution.

Next, remove the Mylar positive from the board and place it in the developer with the copper side up. Gently rock the tray back and forth for 1 to 2 minutes while the board develops. At 68 to 70 degrees Fahrenheit, the normal developing time is 1 to 1½ minutes using fresh developer. As the developer ages through use, more developing time will be required.

A quick check can be made to see if the board has developed properly by dipping it in the ferric chloride etching solution for 2 to 3 seconds and then rinsing it gently with cool tap water. If there is still some unwanted sensitizer that protects some areas, return the PCB to the developer for another 20 seconds or so while gently agitating the liquid solution.

Drying the Developed Board. After the board has been developed in the developer solution, rinse it in cool tap water to remove all developer and let it stand vertically to dry. You can use an ordinary electric fan to help speed up the drying time. When the board is completely dry, it is ready for etching in an etching solution.

Etching the Board. The processes completed so far are similar to photographic work. The board now has a "picture" of the PC circuit "drawn" or "printed" on it so the islands and traces will be permanently left on the board in the form of copper and the rest of the unwanted copper will be etched, or dissolved, away by a solution of ferric chloride (a weak acid solution).

Immerse the board in a separate glass or plastic tray that is filled with about 6 ounces of etching solution. It normally requires about 6 ounces of etching solution for a one-sided 4 x 6-inch board. Make sure the board is completely covered by the liquid with the copper side up. Gently rock the tray back and forth every so often. The etch time varies from 20 to 60 minutes depending on how often you agitate the tray and the temperature of the solution. You can accelerate the etching by warming the solution to between 90 and 120 degrees Fahrenheit. However, if you do warm the solution, be sure to have adequate ventilation.

After the board has been etched for the recommended time, remove it from the solution and rinse it under tap water. Examine the board to see if you are satisfied that all unwanted copper has been etched away and removed. If not, return the board to the etching solution for additional time and then rinse again in tap water.

Removing the Resist. One final step is necessary before your PCB is ready for "parts stuffing." You must remove the etch resist from the copper circuit that remains on the board by using a stripping solution. This solution is available as GC No. 22-240 Stripping Solution, or GC No. 22-209 Print Kote solvent. Or, you can remove the resist using fine steel wool and a little elbow rubbing. Rinse the board in tap water to remove all residue and dry it thoroughly before continuing.

COMMERCIAL BOARD KITS

There are a number of commercially available kits that have all the materials you need to lay out and etch your own boards.

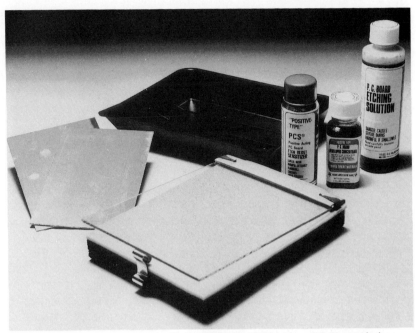

Fig. 7-8. Items that are supplied with the PCS Lab Kit (Courtesy GC Electronics).

PCS Lab Kit

One such kit available from GC Electronics is the PCS Lab Kit. This kit includes everything you need to lay out the positive art, sensitize the copper-clad board, and develop the board. All you need is a UV light source for exposing the sensitized board. The UV source can be a sunlamp or carbon arc.

The PCS Lab Kit is GC No. 22-309 and contains the following (see Fig. 7-8):

➼ Positive Art Kit (includes drafting symbols, tape grid paper, etc.)
➼ Developer concentrate
➼ 4 × 6-inch copper-clad boards (2)
➼ Exposing Frame
➼ PCS Aerosol Positive-Acting Sensitizer
➼ Developing/Etching Trays (2)
➼ Etching Solution (ferric chloride)

P.T.S. Positive Transfer System

The P.T.S. system requires a "negative" transparency of your PCB artwork, though it is called a positive transfer system. A positive transparency means that your circuit layout diagram wiring is opaque (black to light) and a negative transparency is just the opposite (the wiring is white and exposes the wiring when it is printed). In other words, a negative transparency is clear, or translucent (will pass light) wherever your circuit wiring will be.

Before looking in detail as to how the P.T.S. system works, consider in general what the technique does for you when you want a fast system for making only a few copies of a prototype. When you have finished processing the P.T.S. film, you will have a "rub-on decal" that can be applied to a copper PCB by burnishing (rubbing).

In the event you are beginning your PC effort with a positive transparency, you will need a negative transparency produced by a photographer or by making your own, using GC *Reversing Film*. The reversing film procedure is a simple process you can use to process your own negative film. A complete set of GC Printed Circuit Instruction/Information Sheets is available in a packet as GC No. 22-101-HB.

Exposing the P.T.S. Film. As the P.T.S. film is semi-photosensitive in normal room light, it should not be kept out of its light-safe envelope for longer than 5 minutes. The less the premature exposure to light, the better the result.

First, cut a piece of P.T.S. film to the size of the negative circuit layout. The film should have the brown emulsion down. Center the negative on top of the P.T.S. film and cover them both with a sheet of glass to keep them flat. Place the glass, negative, and film under an ultraviolet light from a 275-watt sunlamp. The lamp should be about 12 inches from the film. Ultraviolet light output might vary from lamp to lamp, but the exposure should complete in about 1½ minutes.

Developing. After the film has been exposed, peel off the protective clear plastic coating with the brown side of the P.T.S. film up. Lay the film down on a clean, flat surface (with the brown side still up) and pour a small amount of P.T.S. developer (GC No. 22-216 or 22-217) onto the film to completely wet the surface. The developer should be spread evenly over the surface of the film with the applicator pad that is supplied with the developer. Then we let the film sit for 10 seconds.

Next, wipe the film surface with a non-woven, disposable developing pad (GS No. 22-214) to remove the P.T.S. film emulsion. Use a circular motion and light pressure so the coating is removed by the rubbing action, leaving the desired image. It might be necessary to add small amounts of developer to maintain complete wetting of the film and the applicator pad.

In the event the image rubs off or is not as sharp as the original circuit, the film might not be exposed correctly. Trial exposure times might be necessary, because the exposure will be affected by the opaqueness of the original artwork and changes in efficiency of the light source.

Applying Transparency to PCB. The board should be clean and free of oxides and trace oils from handling. You can clean the board with a household cleaner and very fine steel wool as discussed earlier. Dry the board thoroughly before applying the P.T.S. negative.

Position the P.T.S. transparency on the board with the emulsion (dull) side down. After that, rub the entire image onto the board, burnishing with a flat, blunt instrument such as a wooden popsicle stick. Then carefully peel off the plastic film so the circuit pattern remains on the board as an etch-resistant black coating.

Etching the Board. Etch the circuit on the board by placing it in a glass or plastic tray of ferric chloride etching solution. Remember to agitate the solution by gently rocking the tray back and forth until all excess copper plating has been etched away. When etching is complete, use PCB stripping solution (GC No. 22-240) or fine steel wool and cleanser to remove the P.T.S. emulsion. Your circuit will remain on the PCB, ready for drilling and component parts stuffing.

Radio Shack PCB Kit

Another kit for etching your own PCB is the PCB Kit available from Radio Shack stores (No. 276-1530). See Fig. 7-9. This kit includes the following items:

- 4½ by 3-inch copper-clad boards (2)
- Resist-ink pen
- Resist-ink solvent
- Etchant
- Layout strips and circles
- Developing tank
- ⅛₆-inch drill bit
- Instructions for use.

DIRECT-ETCH PCB FOR SIMPLE CIRCUITS

The direct-etch method of making PCBs is appealing because it is so very economical and requires little time to make a board. It is especially useful when you have a simple circuit that has a few components and relatively few conductors to be etched. The direct-etch method is the easiest method to understand and complete.

Fig. 7-9. PCB kit comes complete with all items necessary for etching your own PCB (Courtesy Radio Shack, Division of Tandy Corp.).

Preparing the Board

Begin with a copper clad board. With a fine-point pen containing special ink, simply draw the conductors and donuts directly on the copper. Any large ground islands are simply "colored-in." The ink dries instantly and is easy to handle. If you need to "erase" an area, just use the stripping solution. Any copper that is properly protected to resist the etching acid will remain behind and become the wiring for your circuit.

Etching the Board

Use about 6 ounces of solution in a plastic or glass tray so the board is completely covered. Again, place the board in the tray and agitate the solution to speed the etching process. Remember that faster etching is better because the acid can undercut your etch resist if too much time is required to do the job.

Cleanup

Rinse the etched board in tap water after all undesired copper is etched away. Any remaining ink and lacquer can be removed with stripping solution. Complete the job with a final polishing using a fine steel wool so that the board will be easier to solder to. Then drill holes and stuff with components.

Etch resist pens are available from GC Electronics as GC No. 22-220 (1/32 inches wide) or GC No. 22-222 (Fine Line 1/64 inches wide). A pen is also available from Radio Shack as Catalog No. 276-1530, and etching solution is available as No. 276-1535.

TEC-200 IMAGE FILM

The TEC-200 Image Film is a new, easy way of making printed circuit boards. It avoids the use of resist pens or rub-on transfers and uses an ordinary plain-paper copier that uses a toner. The simple steps for use of the TEC-200 Image Film are outlined below.

Photocopying the Circuit

The desired circuit pattern from any electronic magazine or your own layout diagram is photocopied onto the TEC-200 film. You can use any photocopier that uses a toner and heat fusing by simply placing the standard size sheet of film in the paper tray, either side up.

Pattern Transfer

After having copied the pattern onto the film, cut the pattern out of the sheet. Leave about a half-inch border of clear film around the edge of the pattern. Next, place the film with the pattern side down onto the copper-clad board. Then, transfer the circuit pattern from the film directly onto the copper by pressing it with a hot pressing iron (not a soldering iron!). The temperature of the iron should be set to the "cotton/linen" setting, or to 265 to 290 degrees Fahrenheit. You can use a thin cotton or muslin pressing cloth on top of the film before pressing, if desired. The TEC-200 aids the copy material (the toner) to melt and form a varnish-like, acid-proof coating on the PCB that is now the etch resist of the circuit. Figure 7-11 shows the TEC-200 film being peeled from the copper board, the circuit diagram remaining.

Etching the Circuit

The PCB can be etched as soon as it has cooled from the ironing process. First, slowly and carefully peel the film off the PCB. The circuit pattern should remain on the board. The board can then be placed into the etching solution where it is etched, following the procedures outlined earlier. After etching the board, cleanse it thoroughly with tap water. The remaining etch material can be removed by washing the

Fig. 7-10. TEC-200 Image Film uses a plain-paper copier to copy a circuit diagram into the film, which is then hot-iron pressed on the copper PCB (Courtesy of the Meadowlake Corp.).

board with nail polish remover or paint thinner. Again, rinse thoroughly with tap water. The board is now ready for hole-drilling and parts-stuffing.

TEC-200 Image Film is distributed by The Meadowlake Corp., 25 Blanchard Drive, P.O. Box 497, Northport, NY 11768.

DASH-PCB LAYOUT

Shown in Fig. 7-11 is the DASH-PCB by FutureNet that uses an IBM PC as an innovative PCB layout "system expert." Circuit layout tasks can be completed four to six times faster than when done manually. The system is not solely for computer enthusiasts or hobbyists, as it is meant to be used by companies large enough to require such "expert" aid in laying out their PCB designs. The DASH-PCB shows how the power of the personal computer can be used to design and rearrange your circuits. Size can be as large as 32 x 32 inches.

Information on the DASH-PCB is available from FutureNet Corp., 9310 Topanga Canyon Blvd., Chatsworth, CA 91311-5728.

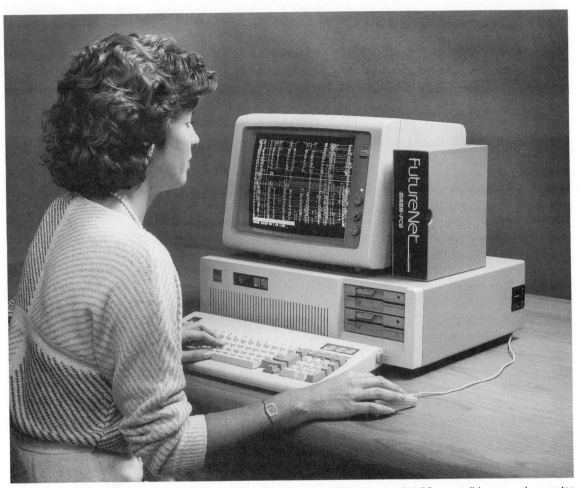

Fig. 7-11. The DASH-PCB is a software program that lets you lay out a PCB using an IBM-PC compatible personal computer (Courtesy Future Net Corp.).

DRILLING THE BOARD

After you have completed the etching of your board, you now need to drill holes in the donuts where component leads will be inserted. As shown in Fig. 7-12, the drill size must be slightly larger than the size of the wire lead to be inserted in the donut pad. However, it must also be small enough so that it does not drill out all the conductive copper that has been so carefully etched onto the board.

The hole and pad diameter depend on each other. To determine the hole diameter, you need a minimum of 0.006 inches to a maximum of 0.020 inches over the lead diameter. The minimum hole size diameter

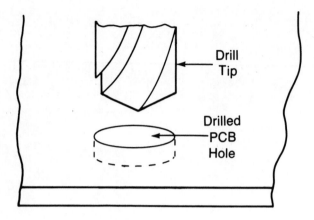

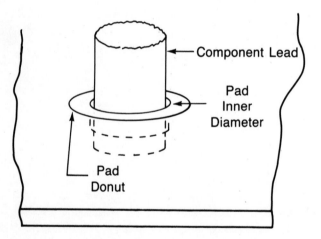

Fig. 7-12. Drilling the donut pad to accommodate component leads.

is then the largest of the two, as shown in Fig. 7-13. If a wire lead is 0.021 inches maximum diameter, then the pad hole diameter must be 0.021 plus 0.020 or 0.041 inches. The maximum pad diameter would then be figured by allowing a minimum of a 0.020-inch ring over the maximum hole size. Figure 7-14 is an example of how the pad size is calculated.

MOUNTING AND SOLDERING COMPONENTS

You are now ready to stuff the components to complete your circuit. Gently bend resistor, capacitor, and transistor leads until they fit easily

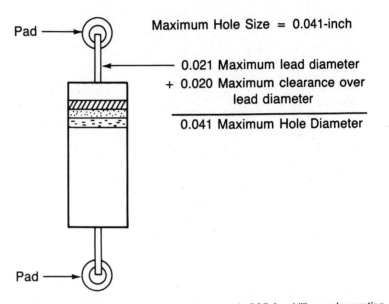

Maximum Hole Size = 0.041-inch

0.021 Maximum lead diameter
+ 0.020 Maximum clearance over lead diameter
_____
0.041 Maximum Hole Diameter

Fig. 7-13. Example of maximum hole size diameter in PCB for drilling and mounting component.

into the holes you have drilled. Refer to Chapter 6 under the section "Component Mounting" for additional reminders. The leads can be bent in or out slightly to hold the component to the board before soldering, as shown in Fig. 7-15. After soldering (see Chapter 6), the wire leads should be trimmed with a pair of cutters.

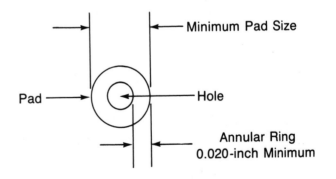

0.041 Maximum Hole Diameter
+ 0.040 Minimum Annular Ring × 2 = 0.040 inches
_____
0.081-inch Minimum Pad Size to Use

Fig. 7-14. Example of how a pad size is chosen.

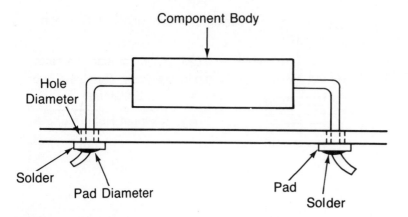

Fig. 7-15. *Insert parts in donut pads, bend leads slightly, apply solder, and cut excess lead wire.*

BOARD CLEANUP AND TESTING

After you have completed inserting all components and soldering them into position, check the entire circuit and components again. Remove any excessive solder rosin that might be on the printed circuit wiring. Make a quick continuity check of each soldered lead to see if it is good. Wiggle the leads to see if they are soldered firmly. An audio continuity tester is helpful in that you can "hear" the trouble without having to look back and forth at the meter.

It is best to do some simple tests on your board before applying power to it. One of these is to see that there are no solder bridges between the various printed conductors or component leads. Use a small magnifying glass to aid you. If any of the printed wiring appears broken, apply a small amount of solder with a heated iron to complete any hairline cracks.

Check the ohmic value from critical supply voltages to ground to make sure you don't have a short across the power supply when turned on. Check the input and output leads for integrity and check all component values to verify you have inserted the correct leads into the correct pads. This easy pre-troubleshooting is all done without turning on any voltages. You might locate an improper part or connection and save yourself a lot of trouble.

The final test is to connect all power supply voltages and give the board its first "smoke test." Have a voltmeter across the power supply leads to verify there is a voltage drop from the load of the circuit. If things look well, congratulations—you have a winner!

CHAPTER 8

Commercial
Assembly Kits

There are a number of commercially available electronic kits that can be assembled by the beginner in electronics or the advanced engineer or scientist who has just begun to dabble in "homemade" electronic projects. This chapter discusses some of the project kits that are available from local electronics stores, by mail, or other delivery services. Most of the kits are very easy to assemble in just a few hours or days and are available at a modest cost. The projects covered are only a few examples of the many available. Consult electronics magazines, amateur radio periodicals, citizen band magazines, etc. for additional projects and prices.

ELECTRONICS KITS INTERNATIONAL

Electronics Kits International (EKI) has a number of build-it-yourself kits that cover the wide gamut of electronic circuit assembly. The circuits are easy to assemble and all come complete with detailed information and a step-by-step procedure as to how to mount the parts. The EKI kits, most of which are supplied with a parts kit and circuit board, are provided with printed instructions, drawings, and photographs. Information provided includes:

➤ project description
➤ step-by-step construction diagram
➤ schematic diagram and circuit board
➤ parts list
➤ theory of circuit operation
➤ pictorial diagram
➤ troubleshooting instructions·

The EKI parts kits, available in many electronics stores, are listed here with kit number and project description.

➤ 801—Pocket Dice
A portable, 9-volt-battery-operated dual pocket dice. You can use it with any game of chance.

➤ 802—Signal Injector
An important and fundamental piece of test equipment that belongs on any serious technicians bench for other projects requiring a signal generator.

➤ 803—Space War Gun
Can be placed in a "rapid-fire" or "stun" mode. LEDs blink, and a loudspeaker transmits high-pitched tones.

➤ 804—Metal Detector
With this kit you can find new money and old treasures. Get started in the fascinating hobby of metal detecting.

➤ 805—Logic Probe
The perfect low-cost logic probe that works with any logic family as it indicates a HIGH, LOW, or PULSE condition.

➤ 806—Burglar Alarm
A great all-purpose, battery-powered burglar alarm with normally open and normally closed contacts. Comes complete with a 9-volt buzzer.

➤ 808—Decision Maker
A super heads-tails or yes-no game. Don't make decisions without it.

➤ 809—LED Pendulum Metronome
Featured in Popular Electronics, the Pendulum Metronome uses an arc of LEDs and a speaker for a unique and exciting display. This is great for any music buff.

➤ 811—Double Decision Maker

An LED decision maker with a "second opinion." Makes for an inexpensive and unusual gift.

➤ 812—Siren Oscillator

As a siren or code-practice oscillator, the 812 is a great sound-effects project. This is a perfect starter kit.

➤ 814—Robot Blinker

Everyone is into robots today. Build your own flashing LED Robot Blinker and amaze your friends.

➤ 816—Mini-Wink

An exciting project that flashes six neon bulbs in a totally random manner. Everyone will wonder what it is.

➤ 818—Fish Caller

Makes a clicking sound in the water to attract fish. Place in a plastic bag or glass jar, tie it to a string, and throw it in the water. You're ready to reel them in.

➤ 820—Shimmer Lights

Causes a 120-volt light to "shimmer." Great as a window display, holiday decoration, or exciting attention-getter.

➤ 821—Christmas Tree

This 9-volt Christmas Tree flashes eight LEDs at random. Hang it on a real tree or place it in a self-standing mode.

➤ 822—One-Channel Color Organ

Flashes a string of lights to the sound of music. Connects to any FM radio, stereo, phonograph, or tape player.

➤ 824—Automatic Siren

An automatic siren that sounds like the real thing. Use it with a burglar alarm, as an attention getter, or for just plain fun.

➤ 826—Fuzz Box

A two-transformer amplifier that adds "distortion" (clipping) to the output signal of an electric guitar. Operates on a 9-volt battery.

➤ 828—0-to 9-volt Power Supply

Supplies 9 volts dc to a host of projects from portable radios to burglar alarms. A perfect "battery eliminator."

➤ 830—Power Supply (Regulated)

A reliable, regulated test-bench power supply adjustable from 0 to 15 volts, 300 mA maximum.

➤ 834—Color Organ (3 Channels, 1 Control)

The 3-channel color organ gives a dramatic effect by flashing three strings of lights to the sound of music from your FM radio, stereo receiver, or high-fi amplifier. Up to 150 watts incandescent per channel.

➤ 836—Color Organ (3 Channels, 4 Controls)

This color organ provides individual sensitivity controls for each channel. Up to 150 watts incandescent per channel.

➤ 840—Variable Strobe Light

A super strobe that produces a very high intensity white light for short durations. Adjustable flash rate.

➤ 842—12-Volt Color Organ

A great 3-channel color organ for use with 12-volt systems such as those found in automobiles. Each channel responds to a particular range of audio frequencies (low, middle, or high).

➤ 844—TV Jammer

Wipes out audio and video TV signals. Operates from up to 40 feet away.

➤ 846—Strobe Stick

An unusual strobe light that can be used as an emergency flasher for automobiles, boats, or campers. Plugs into the vehicle's cigarette lighter.

➤ 855—Whopper Alarm

Makes for an excellent alarm for the 852 Combination Lock/Alarm Control. Can also be used independently as a siren device.

➤ 852—Combination Lock/Alarm Control

A well-designed and sophisticated alarm control system. For use in the home or car.

➤ 856—Electronic Tennis Play

Play fast-action doubles with just one finger. Designed for two players.

➤ 858—Digital roulette

Be a sure winner with an LED Digital Roulette. Operates on four 1 ½-volt D cells. Uses 33 LEDs.

➤ 860—5-to 24-Volt Regulated Power Supply

The perfect power supply for experimenting, project-building, and electronics repair. Provides a fully regulated voltage at up to 1 ampere over its range of 5 to 24 volts.

➡ 861—Big Sound Portable Organ

Far more than a sound generator—a music maker. Produces four octaves and 29 precisely determined notes.

➡ 862—Full-Wave Motor Speed Control

Can be used to vary the speed of a hand drill or other small brush-type motor. Also makes an excellent lamp dimmer.

➡ 866—Digital Slot Machine

Las Vegas and Atlantic City watch out—the all-electronic "one-armed bandit" has arrived. Operates on four 1 ½-volt C cells.

➡ 868—Digital Dice

Loads of fun at home or in your car. A great conversation piece.

➡ 870—Nerve Tester

Two projects in one; the 870 is designed to test your ability to remain calm or your "steady-hand" state. Nervousness is "rewarded" with a mild shock *or* by an LED indicator.

➡ 876—6-Digit Digital Clock

An attractive six-digit, 12/24-hour LED clock. Seven-segment displays are ½ inch high.

➡ 878—Digital Bird

Produces sounds from a tweeting canary to a silly chicken, all with the turn of a few potentiometers. A good introduction to sound synthesis.

➡ 880—12-Volt, 2 Amp Regulated Power Supply

Produces a regulated 13.8 volts dc at 2 to 3 amperes. Ideal for powering CB radios, car-stereos, amateur 2-meter transceivers, or any other 12-volt automotive electronic accessory.

➡ 882—Musical Horn

Designed to be used in your home, car, or boat. Comes with two song IC chips—La Cucaracha and Dixie.

➡ 884—Sound-Activated Color Organ (1 Channel)

"Speak and the lights shall flicker." Use as an applause meter, table display, or sound alarm.

➡ 886—Audio Amplifier/Intercom

Use as a public address (PA) amplifier, phono amplifier, a "bug," or intercom. Requires an additional 8-ohm speaker for use as an intercom.

➤ 888—Librarian Tormentor

A fun game of hide and seek. Produces a high pitch for a few seconds, shuts down for a couple of minutes, then repeats over and over.

➤ 890—Stop-action Timing Tester

Test and expand your sense of timing. "Freeze" the LED by pressing the pushbutton at exactly the right time.

➤ 892—Telephone Hold Button

Put all incoming or outgoing calls on hold. Works with a single phone or any number of extensions.

➤ 894—Phasor Gun/Sound Generator

A super phasor gun and complex sound generator. Change just a few external components and produce all kinds of fascinating sounds.

➤ 898—Binary Clock

Only if you know something about the binary number system will you be able to tell time with this project. A truly intriguing timepiece.

➤ 899—Body Blinker

An excellent first project. For bike riding, jogging, or walking at night. Provides a light to warn motorists of your presence.

➤ 466—EKI Digital Trainer

Learn to analyze, design, and troubleshoot digital circuits by building the EKI digital trainer. Perform many experiments by studying the course lessons.

➤ 467—Microprocessor Course

This complete microprocessor training package includes a microprocessor training manual, a parts kit to do experiments with, and the EKI Micro-Mentor system kit.

➤ 471—Introduction to Electronic Components and Circuit Assembly Course.

This course includes training manuals, breadboard, parts kits, circuit board design projects, and EKI project kits.

➤ 472—AC-DC The Fundamentals of Power Supplies Course

This course is complete with training manual, parts kit, power supply, and six EKI project kits.

➤ 473—Amplifiers and Oscillators Course

A training manual, activity parts kit, audio amplifier project and six EKI kits are included in this course.

➥ 474—Digital Electronics and Integrated Circuits Course

Complete with training manual and activity parts kit, this course has two projects and six EKI kits included.

➥ Analog Electronics (AM Radio Course)

This course has a training manual with lessons and activities, a parts activity kit (breadboard material for AM radio), an AM radio parts project, and six EKI kits.

➥ 476—Advanced Digital and Operational Amplifiers (Digital Trainer Course)

With advancing technology, this course has a training manual, electronic components for experimentation, a digital trainer, and six EKI kits.

➥ 477—Microprocessor and Machine Language Programming Course

The most advanced course, 477 includes a training manual with lessons and activities, electronic components for experimentation, a Micro-Mentor system, and four advanced EKI kits.

To order any of the kits and courses outlined above, or to locate the nearest store, contact Electronic Kits International, Inc., 23210 Del Lago Drive, Laguna Hills, CA 92653, or call, in California, 1-714-859-1948, and out of state, 1-800-453-1708.

RADIOKIT

Radiokit provides builders with a PCB and parts that allow them to construct the device they have read about in the various amateur radio journals or magazines, such as *QST, 73 Magazine, Ham Radio*, and *Ham Radio Horizons*.

These kits are available from Radiokit, Box 411, Greenville, NH 03048 (603) 878-1033.

75-Meter Compact SSB Transceiver

This kit features a superheterodyne receiver and transmitter that covers the 3.8 to 4.0 MHz amateur band. Figure 8-1 shows the Radiokit SSB transceiver that comes complete with two PCBs and is housed in an attractive pre-drilled and lettered cabinet. The kit has all parts and an enclosure and measures 2 × 6 × 6 inches when assembled.

OSCAR PSK Demodulator

This kit decodes the amateur-sponsored satellite repeater transmissions that contain information on the satellites housekeeping

Fig. 8-1. Assembled Radiokit 75-Meter single-sideband transceiver with 30 watts output (Courtesy Radiokit).

status reports. It also reports on environmental data it has collected as well as plain text bulletins. It takes the audio output from an SSB receiver or cassette player and outputs data in a RS232-compatible serial bit stream at 1200 baud. The kit is complete with all parts and enclosure.

Audio Oscillator

This audio function-generator kit includes the function-generator IC, precision frequency-determining capacitors, and printed circuit board with instructions. The oscillator covers 10 Hz to 100 kHz.

Infrared Control Link

This project is a single-channel infrared link with relay output. The kit includes the transmitter and receiver that has a range of up to 3 meters (10 feet). Compact, the transmitter measures 20 × 50 mm and the receiver measures 40 × 40 mm.

The Radiokit PCB requirements are simple in that they use no plated-through holes and are single-sided boards to minimize cost and construction complexity.

RADIO SHACK KITS

There are a number of kits available from Radio Shack stores in your neighborhood. Consult your local store or catalog for all the assembly kits available for various age levels and experience.

200-in-1 Electronic Project Kit

This kit has all parts pre-mounted in a molded case with front controls and speaker. All circuits are connected, disconnected, and reconnected using spring-clip connectors. With this kit, you can build a burglar alarm, telegraph, AM broadcast station, digital timer, electronic organ, radios, and many other projects as described in the 108-page project manual. All projects are operated using 6 AA batteries. This kit is catalog No. 28-265.

60-Project Electronic Lab

This kit has pre-mounted transistors, capacitors, resistors, and diodes to let you build computer circuits, a radio, a Morse code system, and magnetic noise detector. Coil-spring connectors allow easy hook-up of the circuits. This unit requires a 9-volt and 2 AA batteries for operation.

DICK SMITH ELECTRONICS

A number of kits that are supplied complete with all parts, materials, and instructions are available from Dick Smith Electronics. Some of these are described below and can be ordered from Dick Smith Electronics, Inc., 390 Convention Way, Redwood City, CA 94063.

Car Alarm MK2

This alarm kit features exit/entry delay and ''no-false'' alarms. It uses the battery ground strap as a sensor to detect when a courtesy light or other electrical load occurs such as when a thief breaks into a vehicle. It is simple but immune from false triggering, such as magnetic induction, clock winding, or other sources. Order Catalog No. K-3253.

Ignition Killer for Cars

This ''Ignition Killer'' for your car or truck will thwart a thief from driving off with your vehicle should one break in and jump-start the engine. Just as a would-be thief begins to drive off, the engine dies. It will restart and run for a few seconds but die again. In desperation, the thief will give up and look for an easier car to heist, assuming the car has engine problems. The circuit can then be disabled so you can drive the auto in a normal manner. The kit is complete with all parts and mounting case. Order Catalog No. K-3255.

General-Purpose Preamplifier

You might need a stereo amplifier module to amplify a magnetic amplifier, a tape recorder, or microphone. The frequency range of this preamplifier has a response that extends well beyond 20 kHz with gains of 40, 55, and 80 dB by means of simple component changes. The circuit uses an LM382 IC that has left and right channel amplifiers. It operates on 10 to 40 volts dc. Order Catalog No. K-3427.

JAMECO ELECTRONICS

Jameco Electronics has more than 20 kits available at electronics suppliers through the country. Several of these are outlined below.

To obtain kits, contact Jameco Electronics, 1355 Shoreway Rd., Belmont, CA 94002. (415) 592-8097.

JE730—Clock Kit

The JE730 Clock Kit has a four-digit LED display with numerals that are 0.375 inches high. The JE730 is shown in Fig. 8-2. The circuit operates on 12 volts ac that is supplied from a transformer. The IC circuit has a holding feature that allows accurate time setting by ''time slipping'' and a sequential flashing colon that lets you know from a distance that the clock is operating properly.

Fig. 8-2. The JE730 Four-Digit Desk Clock Kit can be set to display 12- or 24-hour operation (Courtesy Jameco Electronics).

The circuit can be adjusted for 12- or 24-hour operation of the clock display. Components are mounted on a printed wiring board assembly and the completed kit is mounted in a case that measures 3³⁄₁₆ inches wide by 1¾ inches high by 1⅜ inches deep. Detailed layout and assembly instructions are provided with the kit. Order JE730.

JE2206B—Function Generator Kit

This function generator uses an XR-2206 monolithic function generator IC that provides the engineer, student, or hobbyist with a highly versatile laboratory instrument for waveform generation. The XR-2206 function generator kit provides three basic waveforms—sine, triangle, and square wave. There are four overlapping frequency ranges that give an overall frequency range of 1 Hz to 100 kHz. In each range, the frequency can be varied over a 100:1 tuning range. The sine or triangle output can be varied from 0 to over 6 volts (peak-to-peak) from a 600-ohm source at the output terminals. The generator has less than 3 percent distortion from 1 Hz to 100 kHz. Order JE2206B.

JE300—Digital Thermometer Kit—Dual Sensors

The dual sensors of this kit allow you to switch and measure both outside and inside temperature from − 40 to 199 degrees Fahrenheit or − 40 to 100 degrees Celsius. Continuous temperature reading is provided by three bright 0.80-inch-high LED digit displays. The thermometer can be switched to display either Fahrenheit or Celsius degrees. The JE300 is shown in Fig. 8-3. The display operates off 120 volts ac and is mounted in a supplied case that measures 3½ inches high, 6⅝ inches wide and 1⅜ inches deep.

Fig. 8-3. Both indoor and outdoor temperature can be measured by the JE300 Digital Thermometer Kit (Courtesy Jameco Electronics).

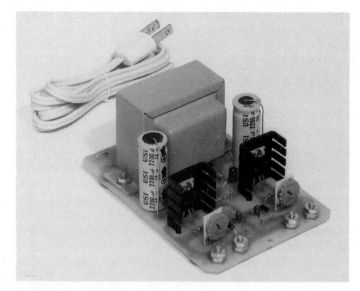

Fig. 8-4. The JE215 Adjustable Power Supply Kit provides both positive and negative voltage supplies (Courtesy Jameco Electronics).

JE215—Adjustable Power Supply Kit

This adjustable power supply kit provides ± 5 to ± 15 volts dc regulated, with a current capacity of 175 to 750 mA per supply. The power supply is shown in Fig. 8-4. The input is 120 volts ac and the unit has adjustable regulators with terminal overload protection.

JE755—Sound Experimenter Kit

There are 76 pieces in this kit which allows you to synthesize sounds such as planes, trains, race cars, birds, whistles, explosions, space warps, phasors, etc. The JE755 is powered by a 9-volt battery or optional wall transformer. The JE755 is shown in Fig. 8-5.

GRAYMARK KOMPONENT KITS

A Komponent Kit is a group of components and instructions that enable the builder to construct a particular circuit. Optional drilled and etched PCBs are available, as are cabinets. The kits described below include assembly instructions, etch patterns, and illustrated parts layout. These kits are nominally priced.

Kits, catalogs, and information can be obtained from Graymark International, Inc., Box 5020, Santa Ana, CA 92704, or by calling toll-free 800-854-7393 (In California, call collect (714) 540-5480.

Fig. 8-5. The JE755 is a sound experimenter kit that has 76 pieces in it. You can synthesize sounds of planes, trains, race cars, birds, whistles, and more (Courtesy Jameco Electronics).

➤ 102—6- or 9-Volt Power Supply
100mA battery eliminator, dual-range output, neon pilot lamp, ac operated.

➤ 103—Mini-Wink Neon Flasher
Random flash pattern, interesting displays, 6 neon lamps, ac operated

➤ 104—Variable Strobe Light
Great for parties and photography, variable flash rate, ac operated.

➤ 105—Fish Caller
Clicking sound imitates distressed fish, variable, battery operated.

➤ 106—Metal Detector Oscillator
Locate hidden metal objects, use with any 9-volt AM radio, battery operated.

➤ 107—Color Organ Control (3 Channels)
200 watts per channel, high/middle/low frequency response, ac operated.

➤ 108—Decision Maker
Makes unbiased decisions (yes or no - not maybe), relaxation oscillator circuit, ac operated.

➤ 109—Auto/Home Burglar Alarm

Use with car horn or Models 110 or 124 sirens, latching circuit, battery operated.

➤ 110—Electronic Whopper Siren

Powerful wailing sound, dual-oscillator circuit, use with any alarm circuit, battery operated.

➤ 112—Proximity Detector

Turn lights on and off, adapt to alarm or chime circuit SCR-controlled, ac operated.

➤ 117—Tunable Electronic Organ

Tunable 7-note scale, play sing-a-long favorites, battery operated.

➤ 119—Motor Speed Control

Adjust motor speed to suit application or use as light dimmer, SCR-controlled, ac operated.

➤ 120—Siren/Code Oscillator

Loud, piercing siren sound or practice Morse code, battery operated.

➤ 122—Computer Sound Effects Generator

Produces weird, spacey sounds, four ICs, tone and rate controls, blip or glide select, battery operated.

➤ 123—Electronic Timer

Turn ac circuit off in prescribed time (2 minutes to 1 hour), one IC, ac operated.

➤ 124—Warbling Siren

Two-tone oscillator siren, loud and piercing, two ICs, use in automobile or with other 12-volt dc system.

➤ 129—Resistance Substitution Kit

Select 24 values from 15 ohms to 10 megohms, with two rotary switches and one slide switch.

➤ 130—Capacitance Substitution Kit

Select 9 values from 0.0001 to 0.22 μF with a rotary switch. All capacitors are rated at 400 volts or higher.

➤ 131—Variable Power Supply

Range is 0 to 15 volts dc, 300 mA, regulated with zener diode and transistor.

➤ 132—Electronic Roulette Wheel
Seven ICs control 38 LEDs and a speaker, providing a decreasing revolution rate and "casino sound."

➤ 133—Logic Probe
Test TTL or CMOS circuits, HI/LO/PULSE readout using three LEDs, 3- to 18-volt dc operating range.

➤ 134—Digital Key
Six pushbuttons allow for more than 46,000 possibilities, timed entry/exit with automatic latching, 5- to 18-volt dc operation.

➤ 135—5-Volt Power Supply
Provides 1.5 amperes at 5 volts dc, IC voltage regulator, ripple is less than 0.5 percent at full load.

➤ 136—Battery Checker
Bargraph LED readout shows condition of 1.5-volt or 9-volt batteries, precision resistance ladder, battery operated.

➤ 137—CFM Decoder
Use with your FM receiver to provide commercial-free music, built-in preamp and FM IF module.

➤ 38—Double Fuzz Box
Hook between amplifier and guitar for expanding sound, select Fuzz 1 or Fuzz 2, includes two foot-switches and gain control, battery operated.

➤ 140—Audio Amplifier
High sensitivity, high gain, use as intercom, PA amplifier, phone pickup and more, two ICs, battery operated.

➤ 141—One-IC Radio
AM broadcast band, high selectivity, no alignment necessary, earphone included, battery operated.

➤ 142—Complex Sound Generator
Create sounds of phasor guns, sirens, train, prop plane, space ship, and more. One IC, 6-position DIP switch, battery operated.

➤ 143—FM Wireless Mike
Broadcast through any FM radio or listen without being seen, high sensitivity, condenser microphone, battery operated.

➤ 144—Transistor/Diode Checker
Test npn or pnp transistors, LEDs, and diodes. Four LEDs indicate short, open, or good condition. One IC, battery operated.

➤ 145—Two-IC Radio

Big AM radio sound from just a few parts, no alignment necessary, speaker included, battery operated.

➤ 146—Digital Dice

Handy, portable set of electronic dice for your favorite games, 14 LEDs for realistic dice count, battery operated.

➤ 147—Sound-Operated Switch

Switch ac on or off with just a clap of your hands, latching circuit with sensitivity control, ac operated.

➤ 148—Headlight Delay

An extra 50 seconds of light after you park your car, two ICs and 12-volt relay, operates on car battery.

➤ 149—Telephone Hold Button

Easily put incoming or outgoing calls on hold, red LED ''on-hold'' indicator, operates on telephone line.

➤ 150—TV Sound Enhancer

''Stereo-like'' sound from your TV, hooks up between TV earphone jack and your stereo system, cables included.

➤ 151—Sound-Activated Color Organ (1 Channel)

Flashing lights from sound waves, no direct connection to sound source, 200 watts, ac operated.

➤ 152—Sound-Activated Color Organ (3 Channels)

Get 600 watts of pulsing lights with no direct connection to sound source, separate control for each channel, ac operated.

➤ 153—Electronic Doorbell

Select from 25 songs plus chimes, plays 8 notes of each timeless favorite, includes two DIP switches, speaker, and special IC. Use with existing transformer.

➤ 154—Happy Face LED Blinker

Put a smile on everyone's face. Happy face PCB comes with kit, IC, LEDs, and connecting wires, battery operated.

➤ 155—Telephone Hold With Music

Provide pleasant listening when you put someone on hold. Connects to earphone jack on your radio or other sound source. Connecting cables supplied.

➤ 156—Musical Car Horn

Select from 25 tunes to announce your arrival, plays 8 notes of each, operates on car battery.

➤ 157—LED Flasher

Random flash pattern of red, green, and yellow LEDs, 8 LEDs included, battery operated.

➤ 158—Radio with Telephone Hold

Radio and telephone hold circuit combined in one kit, includes earphone for tuning and private listening, powered by phone line.

➤ 159—Function Generator

A versatile laboratory instrument at a fraction of the cost of conventional function generators, provides 3 basic waveforms—sine, triangle, and square, 1 Hz to 100 kHz range, ac operated.

➤ 601—Sound-Operated Robot

This speedy little robot keeps banging into walls, but each time it does, it shifts into reverse gear and heads off in a new direction. The instruction manual includes step-by-step assembly instructions with numerous illustrations. Requires two 1.5-volt batteries.

➤ 602—Line Tracer Robot

This robot can "see" and will follow a straight or curved pathway made with a black felt marker or black tape. The infra-red emitter and light sensor circuitry controls a servo feedback loop that causes the motors to make course corrections. This kit demonstrates the principles of electro-optics, analog circuitry, and digital circuitry. The robot operates off one 9-volt and two 1.5-volt AA batteries.

➤ 603—Programmable Robot

You can program this unique robot to go forward, turn right, turn left, pause, sound a buzzer, light an LED, and then repeat the sequence. The robot has a IK static RAM IC that can hold 256 instructions programmed by seven function teach pendants provided with the kit. The robot operates off one 9-volt and two 1.5-volt AA batteries.

➤ 604—Robot Wheel

This super robot is operated by remote control. You build the Robot Wheel and a remote hand-held transmitter control. The two large wheels roll the robot forward, turn it right or left, or you can spin it around. Step-by-step assembly instructions are provided with the kit. The robot operates on two 9-volt and three 1.5 volt AA batteries.

➤ 605—Robot Navigator

This unique robot is programmed to follow a specific course using a black and white paper disc that you program. A photointerrupter circuit reads the rotating disc which causes the robot to go straight, turn right or left, or stop. The kit demonstrates Schmitt trigger circuit operation, optoelectronic circuitry, and motor drive circuitry and control. The robot operates on one 9-volt and two 1.5-volt AA batteries.

➤ 606—Robot Walker

This six-legged robot walks very quickly in a straight line but manages to avoid walking into walls. An infra-red beam circuit can "see" objects in its path and the robot quickly turns out of the way. This kit discusses the theory of operational amplifiers, reflected light-beam circuitry, and motor control circuitry. The robot operates on one 9-volt and four 1.5-volt AA batteries.

HEATHKIT/ZENITH KITS

These Heathkit/Zenith kits are available from the Heath Company, P.O. Box 1288, Benton Harbor, MI 49022, or from local Heath/Zenith Computer and Electronics Centers in various states throughout the U.S. You can order toll free by calling 1-800-253-0570 or locate the store nearest you.

Heathkit/Zenith Printed Circuit Boards Course EI-3234

This printed circuit board course teaches both the direct pattern and photographic etching techniques. It lets you copy PCB layouts from books and magazines or fabricate your own, from almost any schematic.

Figure 8-6 shows the assortment of PC materials you receive as well as two electronic projects. The course includes all necessary materials, components, and hardware to complete the two kits provided. The kits are a photoelectric lamp switch and a touch switch that lets you turn on a light at the touch of a button. The PC materials include an etch-resist pen, rub-on transfers, art tape, etchant, solvent, a positive pattern, clear acetate, photographic developer, and sensitized PCBs. This kit is Catalog No. EI-3134.

Heathkit/Zenith Fish Caller SK-105

The fish caller SK-105 produces a sound that travels through water to attract hungry fish to your area. Figure 8-7 shows the unit that uses a 555 timer IC and operates on a 9-volt battery. Relax and let the fish come to you!

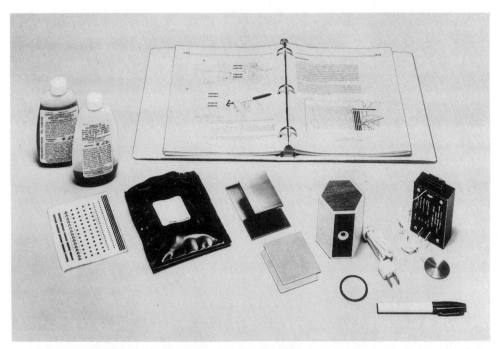

Fig. 8-6. The Heathkit/Zenith EI-3134 Printed Circuit Boards Course includes everything needed. It includes two assembly kits (Courtesy Heath Company, St. Joseph, MI).

Heathkit/Zenith Morse Code Oscillator SK-102

The SK-102 is a Morse code practice oscillator that uses a 555 timer to produce a clear tone that lets you adjust the tone or volume to your liking. Figure 8-8 shows the code practice oscillator, which is good for those studying code for an amateur operator's license.

Fig. 8-7. A simple-to-build fish caller is shown. It uses a 555 timer IC (Courtesy Heath Company, St. Joseph, MI).

Fig. 8-8. The Morse code can be practiced using this easy-to-build circuit that uses a 555 timer IC (Courtesy Heath Company, St. Joseph, MI).

Heathkit/Zenith FM Wireless Microphone SK-106

Figure 8-9 shows an FM wireless microphone that broadcasts through any FM radio up to 50 feet away. The unit is very light and operates on two watch batteries. The wireless microphone can be built in less than one evening. It is housed in a clear plastic housing with ON/OFF switch. In the photo, the left end is the microphone and the right end is the FM transmitter with its flexible antenna wire lead.

Heathkit/Zenith Sound Flasher SK-109

The SK-109 Sound Flasher shown in Fig. 8-10 operates on 120 volts ac and controls and flashes up to three strings of Christmas tree lights. The unit uses a microphone to pick up ambient sounds such as music from a nearby radio or voices in a room to flash tree lights or room lamps.

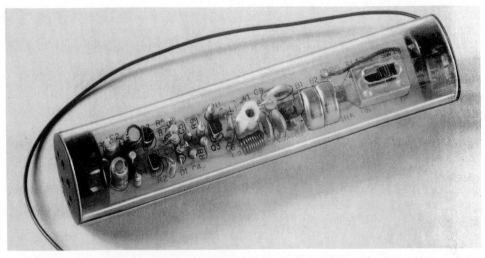

Fig. 8-9. This wireless FM microphone is housed in a clear plastic housing and is powered by two watch batteries (Courtesy Heath Company, St. Joseph, MI).

Fig. 8-10. The SK-109 Sound Flasher can flash Christmas tree lights or a room lamp in sync with music or voices (Courtesy Heath Company, St. Joseph, MI).

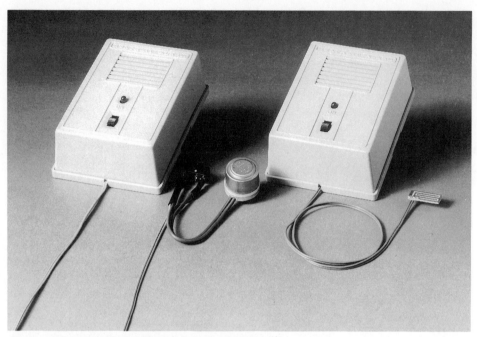

Fig. 8-11. The GD-1183 freezer alarm is shown on the left while the GD-1701 flood alarm is shown on the right (Courtesy Heath Company, St. Joseph, MI).

Heathkit/Zenith Freezer Alarm and Flood Alarm

The Heathkit GD-1183 Freezer Alarm shown on the left in Fig. 8-11 will alarm you if the temperature inside your freezer rises above +20 degrees Fahrenheit. A two-speed beeping alarm and LED will spring into action to warn you if your freezer door is left ajar. The kit can be completed in less than two hours.

Shown on the right in Fig. 8-11 is the GD-1701 Flood Alarm. This unit will alarm you if you have leaking water pipes in the basement or a seeping basement wall. This kit takes about three hours to build and can help you to avoid possible water damage in your home or business.

Heathkit/Zenith Electronic Security Photobeam Alarm

The Heathkit GD-1021 Photobeam Alarm Relay detects intruders instantly when a narrow reflected beam of light is interrupted. The unit is shown in Fig. 8-12 and uses a mirror to reflect a beam of light back to the Photobeam Alarm. The unit operates on 120 volts ac and activates a warning device such as a lamp (up to 300 watts).

Fig. 8-12. The Heathkit GD-1021 Photobeam Alarm Relay detects intruders when a reflected beam of light is broken (Courtesy Heath Company, St. Joseph, MI).

Heathkit/Zenith Surface Mount Technology Kit SMT-1

A surface-mount technology kit is shown in Fig. 8-13. This kit provides valuable hands-on experience with the new surface-mount devices. Special soldering procedures are used for SMT devices, and the kit includes a special tool package and soldering iron.

Heathkit/Zenith Light Meter SMD-1

The SMD-1 is a lightmeter kit and is shown in Fig. 8-14. It makes use of a surface-mount device and is assembled using the SMT-1 kit just described.

149

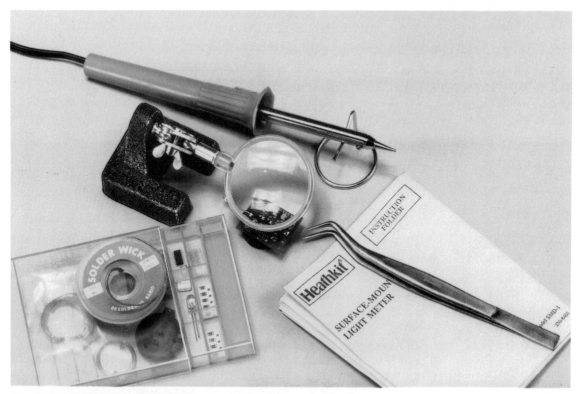

Fig. 8-13. Surface-mount technology is taught through use of the tools supplied as part of the SMT-1 kit (Courtesy Heath Company, St. Joseph, MI).

Fig. 8-14. The Heathkit SMD-1 uses surface-mount technology in this Lightmeter kit (Courtesy Heath Company, St. Joseph, MI).

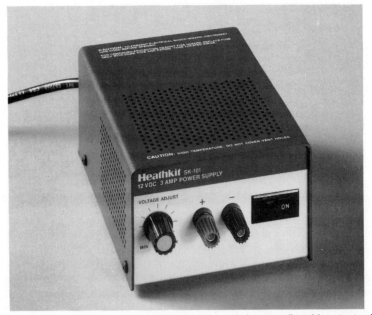

Fig. 8-15. The Heathkit SK-101 12-volt dc power supply has an adjustable output voltage (Courtesy Heath Company, St. Joseph, MI).

Heathkit/Zenith Dc Power Supply SK-101

Shown in Fig. 8-15 is a dc power supply with variable output voltage from 4 to 18 volts dc. Zener diodes also fix the output voltage at either 9 or 14 volts dc to power dc equipment at a current of 2 amperes.

CHAPTER 9

Build These
Practical Projects

This chapter describes a number of simple, practical projects that you can put together in a short time. Some can be completed in as little as a few hours, but others might take several days. You might consider building certain projects first as they serve as test or measurement devices for other, more detailed projects. Exact construction or layout details are left to the builder to determine.

FIELDS OF APPLICATION

The projects to be described can be used in one or more of the following fields. As you gain experience with the use of the projects and understanding of the circuits, you will be able to find other uses for them. When you are finally able to "design" your own circuit or modify one to make it do something different or better, the feeling of satisfaction is great.

Areas where electronic projects can be used are:

➥ Audio Amplification
➥ Automotive
➥ Experimentation
➥ Home
➥ Security
➥ Telephone
➥ Test

152

BRIEF PROJECT DESCRIPTIONS

The following is a list of the projects described in this chapter. I then explain the circuit in detail, give a parts list and suggest other uses for the circuits.

➤ Audio Continuity and Voltage Tester

This very simple circuit uses a piezo sounder to signal the presence of voltage (1 to 24 volts, ac/dc) or to indicate continuity (or low ohmic values) in a circuit.

➤ Continuity/Polarity Tester and LED Power Supply

A step-down transformer provides voltage for operation of this circuit that lets you test for shorts and opens in many electrical devices such as diodes, incandescent lamps, and also provides power to test for polarity of diodes and LEDs.

➤ Intrusion Alarm Trip Circuit

A piezo sounder alerts you to the fact that a door, gate or window has been opened. It trips on an open or short circuit in the alarm protection area.

➤ Noise-Activated Switch and Piezo Sounder

This circuit is activated whenever it "hears" an audio signal or noise above a preset level. A whistle or hand clap will set it off. The circuit then resets automatically.

➤ Annunciator Sounder

This circuit has a piezo sounder that is turned on whenever there is the closure of a switch on a gate, door, or window. The circuit then times out and resets itself.

➤ Whistler Lightning Receiver

By cascading two inexpensive audio amplifiers and with a simple wire-loop antenna, you will be able to listen to *whistlers* that are generated by distant lightning flashes thousands of miles away.

➤ Light Sensor Alarm

When increased light falls on a photocell, this circuit activates a "ding-dong" alarm sounder. At night, it can sense a two-cell flashlight several hundred feet away.

➤ Multiprobe Piezo Sounder

Designed to be activated by many different sensors, this circuit will turn on a "ding-dong" sounder. The input probes can be a water probe, momentary push button, reed relay with magnet, photocell, or others you might think of.

➤ Resistance-Sensitive Audio Continuity Tester

The circuit uses an audio oscillator that changes frequency according to the amount of resistance across its test probes. A loudspeaker produces a high-frequency tone for a low resistance (or a short) and pulses for 10 to 30 megohms resistance. An "open" circuit produces no sound.

➤ Headlight Reminder

As darkness approaches, this circuit alerts a vehicle driver that he should turn on his headlights, or perform other such dusk-dawn activities.

➤ Lamp Dimmer and Motor Control

This is a tried-and-true lamp dimmer circuit that can also control the speed of small 120-volt ac motors such as drills and small saws.

➤ Ultrasonic Transmitter and Receiver

This is an ultrasonic switching system so you can remotely control or activate lights, garage doors, appliances, radios, and TV sets. The turn-on and turn-off distance is about 10 to 20 feet.

➤ Nightlight Turn-on

This is a simple circuit that turns a nightlight on as lights dim and off when light shines brightly on it.

➤ Low-Level Audio Amplifier

When you have a requirement for a small amount of audio gain, this circuit will fit the bill.

➤ Battery Eliminator

You can operate a small radio, tape recorder, or your CB receiver using this basic circuit for a battery eliminator. Input to the unit is 120 volts ac.

➤ Place Telephone on Hold

You can place your telephone on hold and then answer it from any phone in the house with this easy-to-construct circuit.

➤ Visual Telephone Ringer

When there is a lot of noise in a room at home or at a business, you might not be able to hear the phone ring. With this circuit, an incandescent lamp lights each time the phone rings. The flashing light will catch your attention.

AUDIO CONTINUITY AND VOLTAGE TESTER

This handy tester can save you a lot of time and effort in troubleshooting electrical and electronic items at home or in your workshop.

Project Description

Even though it is a very simple circuit, this audio continuity and voltage tester will get a lot of use on your workbench or when working on your car. The voltage tester can be used for testing for the presence of voltage from 1 to 28 volts ac or dc. In the CONTINUITY TEST switch position, it can be used to test for ohmic values from a short (0 ohms) to about 30,000 ohms. The lower the resistance, the louder the tone. The higher the resistance, the lower the volume, and for an open circuit, there is no tone. The circuit draws very little current when being used for voltage sensing and can be used to measure voltages from 1 to 28 volts ac or dc.

Circuit Description

The circuit is shown in Fig. 9-1. The main component of the tester is the piezo sounder that operates off a nominal 1 to 28 volts dc. It can test dc voltages as low as a weak 1.5-volt battery up to 28-volt batteries that are sometimes found on aircraft. As the voltage increases,

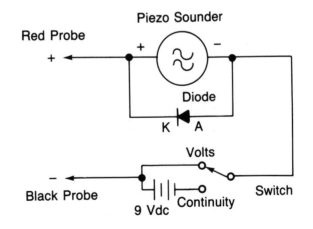
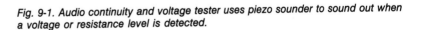

Fig. 9-1. Audio continuity and voltage tester uses piezo sounder to sound out when a voltage or resistance level is detected.

the audio output level increases. Hence, audio intensity is a sign of voltage level.

The diode placed across the piezo sounder is connected in reverse polarity and is in parallel so that the tester will work on ac voltages as well as dc. The positive red probe of the tester is connected to the positive terminal of the sounder. The negative black probe is connected to the negative terminal of the sounder. Correct polarity needs to be observed only when working with dc voltages.

In order to use the tester to check continuity of an inactive circuit, the switch is placed in the CONTINUITY position. This adds a 9-volt battery to the circuit, with the negative of the battery connected to the negative of the sounder. The positive of the battery is connected to the positive of the sounder through the device being tested for continuity. Again, the more resistance in the circuit, the lower the audio tone is. A short at the test probes (no resistance) produces the greatest sound output, which is that of a 9-volt battery connected directly across the piezo sounder. Very little current passes through the circuit under test because the sounder requires very little current to produce a good level of sound output.

As there is very little current drawn when the tester is in operation, the battery should last its shelf life. For many of the batteries available today, especially the alkaline type, you should be able to get up to 5 years life out of a quality battery.

The continuity and voltage tester can be mounted in any kind of housing you like. You can use a plastic pill bottle or container as long as it is large enough. If you use a small camera battery, you can fit all the parts into a 35-mm plastic film container.

The continuity tester is very handy for testing incandescent light bulbs, flashlight bulbs, fuses, transformers, coils, LEDs, diodes, transistors, resistors with values below about 20 kilohms, piezo sounders, capacitors, switches, and just about anything that has some ohmic resistance. The continuity tester is a good device for showing the charge and discharge times of a large capacitor. When you test the capacitor for continuity, the battery in the tester will charge the capacitor to the value of the voltage in the tester (a nominal 9 volts dc). Next, switch the tester to VOLTAGE and again place the tester leads across the large capacitor. Because the capacitor is charged up, it acts as a battery, and the piezo sounder will sound until the capacitor is discharged through the resistance of the sounder. A capacitor of several hundred to one thousand microfarads will cause the sounder to emit a tone for 2 to 5 seconds, a good demonstration of the charge and discharge cycles of a capacitor.

Table 9-1. Parts List for Audio Continuity Tester.

Item	Description
Diode	Diode, 1N4001
PS	Piezo sounder, 4 to 28 Vdc, Radio Shack No. 273-060
SW	Switch, SPDT, Radio Shack No. 275-407 or equivalent
Leads	Test Leads, 30 inches long, Radio Shack No. 278-001, or equivalent
Misc	Case, hookup wire, battery connector

Parts List

The parts list for the continuity tester is shown in Table 9-1. These parts are usually available at local electronics supply houses, but they might be readily available from your electronic "junk box," because none of them are critical in value.

Further Uses of Project

As you gain use with the audio continuity and voltage tester, you will find additional applications for it. For example, when working on a car or truck, the tester can be connected across the back-up lights to verify they are coming on. If so, the sounder will activate when voltage from the lights is applied by placing the car in reverse. In this way, you don't have to get out to see if the backup lights are on, because you can hear the sounder. The same test can be conducted for an oil pressure switch.

CONTINUITY/POLARITY TESTER AND LED POWER SUPPLY

This is another handy test aid that you will want to build for use around the workshop. It is ac-powered, so you can leave it plugged into the wall outlet.

Project Description

The Continuity/Polarity Tester and LED Power Supply is a worthwhile addition to the test devices on your workbench. It operates on

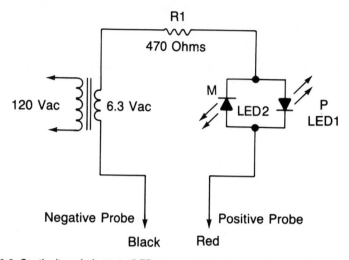

R1
470 Ohms

120 Vac   6.3 Vac

M   LED2   P
LED1

Negative Probe ↓   ↓ Positive Probe

Black   Red

Fig. 9-2. Continuity polarity tester/LED power supply also tests for shorts and opens.

120 volts ac and can be left plugged into a wall outlet because it consumes very little power when in operation.

The heart of the circuit is two LEDs that are connected in reverse polarity and in parallel to indicate direction of current flow (Fig. 9-2). When the two test probes are shorted together, current provided by the transformer flows through both LEDs and they both light. When a resistor is placed across the test probes, the LEDs again light, but they will not be as bright as before because now there is additional resistance in the circuit. This method is used to indicate continuity in a circuit, using the brightness of the LEDs as a measure of resistance across the probes.

When a component such as a diode is placed across the test probes, one of the LEDs will light and the other will not, because the diode is a unidirectional device and current flows in only one direction through it. Which LED that lights depends on which of the test probes is connected to the anode of the diode. In this manner,the tester is used as a polarity tester. It can be used with diodes, LEDs, transistors, transformers, chokes, resistors, earphones, bulbs, lamps, telephones, and virtually any component that has less than about 20 kilohms resistance.

Circuit Description

The circuit for the polarity tester is shown above in Fig. 9-2. Transformer T1 provides 6.3 volts ac, which supplies power for the

158

circuit. R1 is a 470-ohm current-limiting resistor that regulates the amount of current through LED1 and LED2 when the two test leads are shorted. Most LEDs are matched in brightness, but it is helpful if you choose two that do match so that they have the same apparent brightness.

When the circuit is connected as shown, the cathode of LED1 should be connected to the positive or red test probe. This is necessary when using the circuit to test for polarity. With the ac supply voltage, LED1 will light up when the sine wave of voltage is positive on *its* anode, and LED2 will light up when the sine wave of voltage is positive on *its* anode. Half the time both LEDs are on, half the time they are off, alternating with the supply voltage. The negative or black probe is negative with respect to the use of the circuit as a polarity tester.

Testing Bidirectional Devices

Bidirectional components include resistors, light bulbs, photoresistors, transformers, and the like. Bidirectional means current can flow in either direction.

Resistors. A resistor has no polarity by itself. It has polarity of voltage only when current is flowing through it and there is a voltage drop across it. Then, electrons enter the negative side of the resistor and exit the positive side of the resistor.

When you place a resistor across the probes of the tester, current flows in both directions through the resistor, and therefore, both LEDs light up with the same intensity. You can place different values of resistors across the probes and note from the intensity of the LEDs that as the values approach 20 kΩ to 30 kΩ, the intensity begins to dim. This will give you an idea of the relative value of the resistor if it is unmarked. You can judge an unmarked resistor value by making intensity comparisons using known resistor values.

Light Bulbs. You can check continuity of a light bulb by placing it across the test probes. A cold bulb has 5 to 10 ohms of resistance, so both LEDs should light up. This is also true for high-intensity lamps, headlights, and the like. If the bulb is open, neither of the LEDs will light.

Photoresistors. The resistance of a photoresistor (or photocell) varies greatly, from a few hundred ohms to 10 megohms, depending on the amount of light striking it. The greater the light, the lower the resistance. In the dark, they measure 1 to 10 megohms. When the test probes are connected to a photocell, both of the LEDs light with equal intensity. As you cover the photocell with your hand to block out light, the LEDs will dim greatly as the resistance increases up to many megohms. This simple test shows you that the photocell is working properly.

Transformers. Transformers, inductors, solenoids, relay windings, or any item wound with copper wire has a dc resistance of a few tenths of ohms to many tens or hundreds of ohms. When you connect one of these across the test probes, current should flow through the LEDs in both directions so they both light. In the same manner, you can test loudspeakers, earphones, and piezo sounders. When testing these devices, you might hear a 60-Hz hum, which is normal.

Testing Unidirectional Devices

Unidirectional devices include the diode, LED, or solar cell.

Diodes. The simple diode is a unidirectional device, passing more current in one direction than the other. Place a diode across the test probes of the polarity tester. If it is good, one of the LEDs will light brightly and the other LED will be dim or off entirely. This difference in brightness is an indication of the front-to-back ratio of the diode (the degree of how it conducts very well in one direction and not in the other).

Where the polarity tester is especially handy is in identifying the polarity of the diode (which lead is the anode and which is the cathode). Referring once again to Fig. 9-2, LED1 is identified as P (positive) and LED2 is identified as M (minus). The test probes are also labeled as red (positive) and black (negative). If you have connected the diode correctly across the test probes, the red lead is connected to the anode and the "P" LED alone should light. Accordingly, the black test lead (negative) is connected to the negative, or cathode of the diode. Thus, you have rapidly identified the anode and cathode of the diode if you cannot identify the stripe marking that is placed on the cathode ends of the diodes.

If you had "incorrectly" connected the test leads across the diode, the "N" LED would light, telling you that the negative lead is connected to the anode rather than the cathode. Regardless, you have correctly and easily identified the anode and cathode. You can reverse the leads if you want to assure yourself that you have done the test properly.

The three possible conditions of a diode are:

➢ Good diode—Conducts very well in one direction and not in the other direction. One LED on, one LED off. Good front-to-back ratio.
➢ Shorted diode—Conducts equally well in either direction. Both LEDs light. Poor front-to-back ratio. Discard.
➢ Open diode—Conducts poorly in either direction. No LED on. No front-to-back ratio. Discard.

Using the above information, you can also use the tester to check the condition of transistors, because a transistor is really just two diodes back-to-back.

The Solar Cell. When you want to know if a selenium solar cell is good, you can tell its condition by using the polarity tester. The selenium cell has about 1000 to 1500 ohms of resistance, with a poorly defined front-to-back ratio. That is, it has about the same resistance in either direction. However, the amount of resistance it displays is dependent on how much light is striking its surface at the time you are taking measurements.

Both LEDs will light when you place the probes across the leads of the cell (showing there is little front-to-back ratio), but the LED brightness should be somewhat less than when the test leads are shorted together. Therefore, short the test leads together first to get an idea of the relative brightness of the LEDs and then place the leads across the device to be tested. When the device is installed in a circuit, however, be sure there is no voltage applied to the circuit to be tested. Also, be sure you are testing just the device and not some other part or portion of the circuit that is connected across the device.

As An LED Power Supply

The polarity tester also makes an excellent LED power supply and tester. It is a very unique device for doing this because it not only tests LEDs but it serves as a "short-proof" power supply.

If you take an unknown LED with unknown color and polarity and connect it to the test probes, it should light if it is good. Also, one of the tester LEDs should light. If the "P" LED lights, it means you have connected the anode of the LED under test to the positive test probe and thereby identified the leads.

Compare the brightness of the LED under test to that of the LEDs in the tester. Also, you automatically test the two-color or two-current direction LEDs because you are passing an ac current through the LED under test. If you accidentally short out the LED under test, it will go out and both of the tester LEDs will light. Therefore, you don't have to worry about shorting out the tester leads.

Infrared LEDs (IREDs). How do you test an infrared LED when you can't see its emission? Just use the polarity tester. The same rules apply for the IRED as they do for the visible LED. If you have a good front-to-back ratio as indicated by the brightness of the two tester LEDs (one bright, one off or dim), you know you have a good IRED.

Testing Digital LED Displays. You can test the seven-segment LED digital display for proper operation and display the same way you test individual LEDs. Locate the common cathode and decimal point and test the seven-segment anodes for proper display using the polarity tester.

Parts List

The parts list for the continuity and polarity tester is shown in Table 9-2. You can use red or green LEDs for both of the LED indicators, but it is best to not mix them. Use one or the other, because the current drawn by a red and green LED might be different, causing unequal currents and brightness. This difference in brightness could give you a false indication of the front-to-back ratio of the device under test. If you use a transformer with a secondary output voltage other than 6.3 Vac, say 12.6 Vac, double the value of R1, the series current limiting resistor, to 820 or 1000 ohms.

Further Uses of Project

As you use the continuity and polarity tester, you will become familiar with what it can do for you. It is novel because it produces sufficient voltage for use on low-power devices and yet is protected when you short the test leads. Use it to test capacitors, especially those of larger values, because it gives a good indication of the condition and size of the capacitor. As capacitors pass ac, both of the LEDs will light, their brightness depending on the size of the capacitor (a small capacitor will be dimmer than a larger one).

A piezo sounder can be tested, and you will hear a 60-Hz pulsing

Table 9-2. Parts List for Continuity/Polarity Tester.

Item	Description
LED1, LED2	Green or red LEDs
R1	Resistor, 470 ohm, ½ watt
T1	Transformer, primary, 120 Vac; secondary, 6.3 Vac, Radio Shack No. 273-1384, or equivalent
Case	Utility case, Radio Shack No. 270-231, or equivalent
AC cord	Ac line cord, 6 ft.
Misc	Test leads (24 to 36 in.) with alligator clips, hookup wire, solder, screws

sound that is at the frequency of the sounder. The tester is very handy and will not harm any of the devices under test because it is a low-current, low-voltage unit. You might find it takes the place of a voltmeter, a battery, an ohmmeter, or other items you frequently use around your workbench for testing.

INTRUSION ALARM TRIP CIRCUIT

You can build this circuit for the house or office as some protection against intruders. It can also be used to count customers or sound a chime with the arrival of visitors.

Project Description

An intruder alarm system must have some means of sounding an alarm when the system is energized. An intruder alarm is ordinarily used when someone is not on the premises, so an audio alarm needs to be activated to alert nearby homes or offices or the police. The sounding alarm should frighten (or at least unnerve) the intruder into fleeing. If a burglar tries to first cut the wires leading to the siren or bell, this circuit alerts you by sounding the alarm when the wires are cut.

Circuit Description

The alarm trip circuit is shown in Fig. 9-3 and is a fairly simple circuit. The heart of it is the method of obtaining bias for the base circuit of the germanium transistor that is used. The bias-voltage source is included in the intrusion detector. The protected area of the circuit is shown by a dotted circle. Transistor Q1 (a 2N651) has an emitter-to-base voltage drop of a nominal 0.3 volts (because it is a germanium transistor). The base is positive with respect to the emitter. (As a reminder, recall that a silicon transistor has an emitter-to-base voltage drop of a nominal 0.7 volts, again with the base positive with respect to the emitter.)

In the circuit shown, when the connecting wires in the protected area are cut, the bias to transistor Q1 is removed, and it conducts because the base is now biased positive with respect to the emitter. This is the condition for conduction in a transistor—forward bias the emitter-to-base diode and reverse bias the collector diode. When the transistor conducts, the piezo sounder is energized and it produces a "ding-dong" sound. If the alternate connection shown in Fig. 9-3 is used, an alarm bell or siren is set off.

If the intruder should short the wires together in the protected area, this also removes the bias from the base and sets off the alarm. When the protected area leads are shorted, the alarm sounds because of

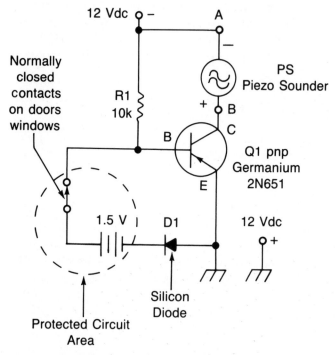

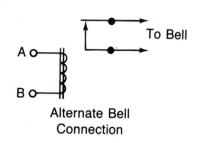

Fig. 9-3. Alarm circuit trips on an open or short.

Table 9-3. Parts List for Alarm Trip Circuit.

Item	Description
D1	Diode, silicon, 1N4001 or equiv.
PS	Piezo sounder, Chime, 6 to 18 Vdc, Radio Shack No. 273-071 or equivalent
Q1	Transistor, 2N651, germanium, Radio Shack No. 276-2007, or equivalent
R1	Resistor, 10k, ½ watt
SW	Switch, SPST, normally closed
Misc	Utility case, battery connectors, batteries, perimeter wiring and switch

the voltage drop across diode D1. The circuit thus trips on an open circuit or a short circuit.

Parts List

Table 9-3 shows the parts list for the alarm trip circuit. The parts are readily available from local electronics supply stores or houses.

Further Uses of Project

The protected area could have its contact switch at a door to a store so that every time someone entered, it would break the circuit and sound the sounder or advance an electromechanical counter. In this manner, you can count the number of visitors per unit time and alert the staff that a customer has entered the premises. This circuit can be used to provide security by alerting you to a sound that has occurred in another part of the house, office, or building.

NOISE-ACTIVATED SWITCH AND PIEZO SOUNDER

This project lets you experiment with a circuit that is activated by a whistle or hand clap.

Project Description

The center of this circuit is an SCR that is triggered into conduction by an amplified hand clap or whistle. It then provides voltage to turn on a piezo sounder. After a short while, the circuit times out and the sounder becomes silent, awaiting another sound strong enough to once again trigger the SCR into conduction.

Circuit Description

Referring to the circuit diagram in Fig. 9-4, a microphone picks up audio sounds and feeds them to transistor Q1 (a 2N222), which amplifies the sounds. Resistor R2, a 5 kΩ potentiometer, is a sensitivity control that sets the level at which the silicon-controlled rectifier will be triggered into conduction. When the SCR is triggered on, it completes the circuit that applies voltage to the piezo sounder (PS). The sounder then emits a sound for a period of about 5 seconds and then the SCR is reset, awaiting additional sounds that exceed the threshold level. Note the polarity of the voltage applied across the piezo sounder. The positive of the sounder goes to the positive of the power supply (pin 3), while the negative is connected to pin 2. Pin 1 is the negative of the 6-volt power supply.

Parts List

Table 9-4 shows the parts list for the noise-activated switch and piezo sounder. The sounder is rated at 4 to 28 volts dc and will operate from about 1 to 30 volts dc. All components are standard. The microphone is an inexpensive one because audio quality is not a consideration. A spare transistor radio earphone can serve adequately in this circuit.

Further Uses of Project

The sounder can be replaced with a 6-volt bulb to provide a visual indication that the circuit has been activated. If the unit is placed near a window or door, it will signal to someone, perhaps an intruder, that their presence has been detected by the sound of breaking glass or of forced entry at a door or window.

ANNUNCIATOR SOUNDER

The annunciator sounder is a handy way of letting you know that someone has opened a gate, door, or window by causing a piezo sounder to ring several times and then go silent. This project can be used for home, office, or store security. It can also serve as a test bed for other applications that you might devise as you become familiar with its merits. It, therefore, also serves as an experimentation project.

Project Description

This circuit has two parts to provide energy to a piezo sounder to alert you that someone has caused circuit closure on a gate, door, or window at some distance from the sounder itself. A momentary "make" switch initiates the operation, times out, and then resets itself. The sounder will then cease sounding.

Circuit Description

The heart of the circuit is the npn transistor Q1 (a 2N2926), which causes the piezo sounder PS to sound (refer to Fig. 9-5). When switch SW, mounted on a gate, door, or window, is closed momentarily, capacitor C1, a 100-microfarad electrolytic, is charged rapidly to about 0.7 volts. At this time, the transistor is forward biased and collector current will flow. This causes the piezo sounder to operate, producing a "ding-dong, ding-dong" sound.

After the switch is released (it only has to be closed momentarily), the 100 μF capacitor begins to discharge through the 33 kΩ resistor, R2, at the base of the transistor. As the voltage at the base begins to

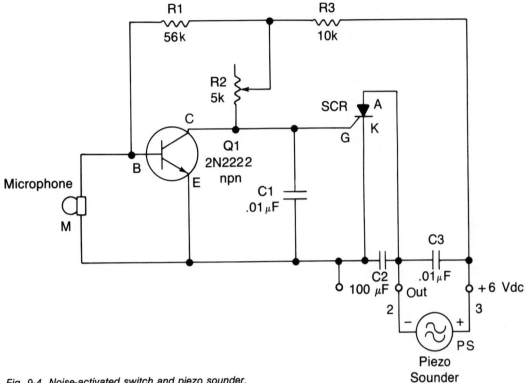

Fig. 9-4. Noise-activated switch and piezo sounder.

Table 9-4. Parts List for Noise-Activated Switch.

Item	Description
C1, C3	Capacitor, 0.01 μF, 35 Vdc
C2	Capacitor, 100 μF, 35 Vdc
M	Microphone, small audio
PS	Piezo sounder, 4 to 28 Vdc, Radio Shack No. 273-060
Q1	Transistor, 2N2222
R1	Resistor, 56k, ½ watt
R2	Potentiometer, 5k
R3	Resistor, 10k, ½ watt
SCR	Silicon-controlled rectifier
Misc	Case, hookup wire, battery connector

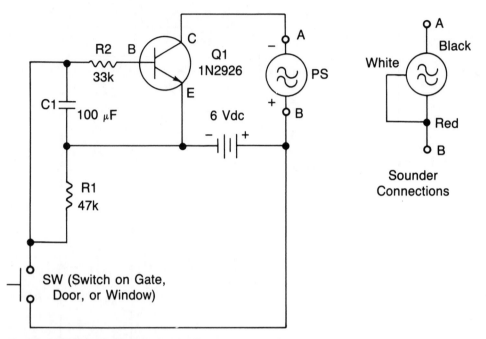

Fig. 9-5. Annunciator Sounder is energized by momentary-make switch placed on gate, door, or window. Circuit then times out and sounder ceases.

drop away from the 0.7 volts, the transistor is no longer forward biased, and it ceases to conduct. When collector current ceases to flow, the piezo sounder shuts off. Capacitor C1 continues to discharge through the 47k resistor, R1. For the values shown, the circuit will remain on for about 8 seconds. If the switch is still closed, the sequence will continue, and the sounder will remain activated.

Parts List

The parts list for the annunciator sounder is shown in Table 9-5 and all parts are readily available at local supply houses. If the dong-dong chime sounder is not available, any piezo sounder or relay-operated bell will do.

Further Uses of Project

While the circuit design calls for a true switch closure in order to initiate operation of the circuit (that is, a short at switch SW), other similar devices can be used to start the action. These include a photocell that can reach a low resistance in bright light, a solar cell that would start the action when exposed to light, a foot switch placed under a carpet,

Table 9-5. Parts List for Annunciator Sounder.

Item	Description
C1	Capacitor, electrolytic, 100 μF, 35 Vdc
R1	Resistor, 47k, ½ watt
R2	Resistor, 33k, ½ watt
Q1	Transistor, npn, 2N2926, Radio Shack No. 276-2009, or equiv.
PS	Piezo sounder, Chime, Radio Shack No. 273-071
SW	Switch, momentary make
Misc	Case, hook-up wire, battery holder and connection

a mercury switch placed on some item that moves when it is used, or the like. Use your imagination!

WHISTLER LIGHTNING RECEIVER

The whistler lightning receiver will give you a listening sensation that you have never before experienced. You can hear the effects of lightning strikes that occur halfway around the earth in your back yard. When conditions are good, you might even hear them from within your house.

Project Description

This project consists of several high-gain audio amplifiers, with a very simple antenna connected to the audio input of one of the amplifiers. You then listen to the audio output of the combination. A radio receiver is configured to be responsive to very low frequencies that fall in the audio range, but the audible signals will be radio frequency and not audio frequency signals. "Sound" signals are those that occur within the audio frequency range of the human ear (from 20 Hz to 20 kHz).

The whistler lightning receiver uses a simple antenna that is about four feet square and uses just a few turns of wire to pick up the very low frequency radio signals sent out by lightning flashes and other sounds within the radio frequency range of 100 Hz to about 15,000 Hz. This is the same frequency range of a high-fidelity audio amplifier.

What you hear in the extremely low frequency radio range can be described as *clicks, pops,* and *tweeks.* Other sounds of atmospherics (static) occur in the early morning and are described as *dawn chorus.* But the main sound this project seeks is the elusive *whistler.* The whistler lasts for 1 to 2 seconds and sweeps across the entire audio frequency spectrum, from the high to the low end. The whistler is believed to result from the dispersion of lightning energy that is propagated through the ionosphere and guided by the lines of force of the earth's magnetic field. Because the higher frequencies travel faster than the lower frequencies, a signal of descending frequency is obtained—a whistle. Other sounds you might hear sound like *hiss, swish,* or a frying sound. They are all very mystifying, indeed!

The whistler lightning receiver falls into the category of audio amplification, because the circuit uses several high-gain audio amplifiers. After you have completed the whistler lightning receiver and have gained some knowledge and experience, you might say you are an ''amateur expert'' in this field.

Circuit Description

Figure 9-6 shows a layout of components that go to make up the system capable of receiving these extremely low frequency signals. The antenna is a four-turn, 4-foot loop that has a miniature output plug.

Audio Amplifier. The loop output serves as the input to the first audio amplifier AM-1. The audio output jack of this amplifier is used to feed the input of the second audio amplifier AM-2. Both amplifiers have small loudspeakers, but only the last one is used to listen to whistler signals. The loudspeaker of AM-1 is disconnected when the cable is inserted into its output jack. The two amplifiers are now cascaded, and the overall gain is that of the first amplifier multiplied by the gain of the second amplifier. The amplifiers are inexpensive and easily obtained from your local electronics store.

The Loop Antenna. The loop antenna is a square loop made by gluing two pieces of 4 × 1½ × 2-inch wood together to form an ''X.'' Four turns of hook-up wire are then wrapped around the ''X.'' The two ends of the loop wire are connected to an audio cable that has a miniature plug on one end. This is the end that is plugged into the first audio amplifier.

Matching Transformer. Additional gain can be obtained by matching the low impedance of the four-turn loop to the input of the audio amplifier by using an audio output transformer, connected backwards. That is, connect the loop antenna to the secondary of the

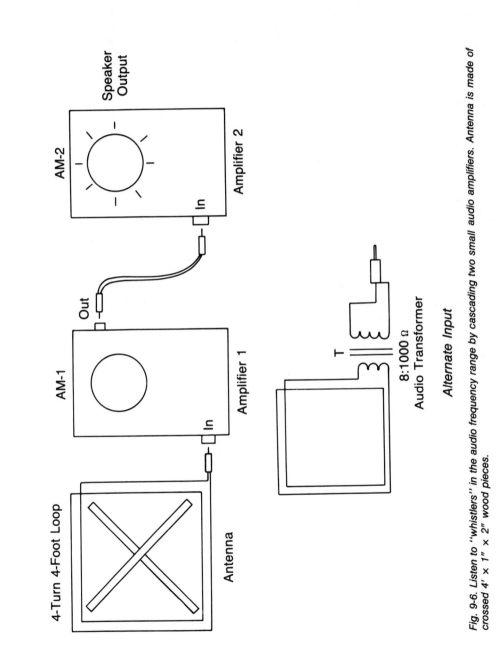

Fig. 9-6. Listen to "whistlers" in the audio frequency range by cascading two small audio amplifiers. Antenna is made of crossed 4' x 1" x 2" wood pieces.

Table 9-6. Parts List for Whistler Lightning Receiver.

Item	Description
Amplifier 1,2	Amplifier, audio, with input and output jacks, speaker, Radio Shack No. 277-1008, or equivalent
Antenna	4-turn, 4-foot square loop
Antenna Cable	Cable, ⅛ inch plug to lug, Radio Shack No. 42-2454 or equiv.
Cable	Audio cable, ⅛ inch mini-plug, male-to-male, Radio Shack No. 42-2420 or equivalent
T	Transformer, audio output, 1000 ohm to 8 ohm, Radio Shack No. 273-1380
Misc	Antenna wire, insulated; wood, 4′ × 1½″ × 2″

transformer which has 8 ohms impedance. The primary of the transformer has 1000 ohms impedance which you now connect to the input of the amplifier.

Parts List

The parts list that it takes to make up the whistler lightning receiver is shown in Table 9-6. Audio amplifiers AM-1 and AM-2 are identical items and provide more than sufficient audio gain to hear the various sounds and effects described. Adjust the audio gain of the first amplifier to a maximum so that it will provide the lowest (best) noise figure for the system. Adjust the audio gain of the last amplifier to a comfortable room level. You can also feed the last amplifier to a larger speaker-amplifier or to a public address system if you desire. You can do the latter if there is not much hum in your area due to house wiring or the like.

The size of the loop antenna is not critical, but the 4-foot wood pieces are a handy size to carry around if traveling farther away from power lines or highway noise.

Further Uses of Project

Patience is necessary when listening for whistlers, because they are elusive. The best time for them is between midnight and dawn. You will be able to hear clicks, pops, and tweeks almost around the clock, because they are so plentiful, especially in higher noise areas. When a local thunderstorm or electrical storm approaches, you will hear hundreds of loud clicks. Listen on a transistor radio with one ear and the whistler receiver with the other. You'll hear the click and the static coincide in time, but they will sound different because of the large difference in the radio frequency. Tune the radio to the upper range of the broadcast (BC) band around 1600 kHz where there are no stations. You'll hear a great amount of static at night, especially with a thunderstorm in the area.

When there is a lightning storm in the local area, do not venture outside with the loop antenna or stand under trees because of the possibility of a lightning strike. Large earth currents can flow through the ground you are standing on and subsequently you if lightning should strike a nearby tree. Stay in the house, barn, or other structure during a storm. Your car is also a safe place, because the vehicle is insulated from the ground by the rubber tires.

When listening to the whistler receiver near a well-traveled road, you might hear unusual sounds caused by passing vehicles, such as the whine of a generator or alternator. Sometimes you can hear the electrical charge that can build up on a car that sounds like the hiss of escaping steam.

When it is electrically quiet and you are listening to your whistler receiver, you might on occasion hear strange noises that come and go, such as an air conditioning unit starting up. You might be able to hear these signals from units that are located several hundred feet away.

If you have an extra-sensitive whistler receiver system, you might hear certain navigation radio signals. The Omega navigation signals operate in a time sequence on frequencies of 10.2 kHz, 11.33 kHz, and 13.6 kHz. The stations transmit for a period of 1 second on each frequency, with the cycle repeating every 10 seconds. These Omega Navigation Systems are located throughout the world at Norway, Liberia, Hawaii, North Dakota, La Reunion Island, Argentina, South Pacific, and Japan. When you hear one of them, you have a sensitive receiver! You will really be thrilled when you hear your first whistler, realizing it has been generated by a lightning flash that occurred half a continent away.

LIGHT SENSOR ALARM

The light sensor alarm is a handy addition to sensors that you might employ around the house or office to provide an intrusion alert or for counting the number of persons visiting your premises. It, therefore, also functions in the security aspects of protecting your area at doors, gates, or windows. By placing the photocell near a floodlight activated by an IR sensor, a sound can be added to that of the light coming on to alert intruders that they have been detected. The light coming on is a silent alarm to the intruder, but adding the sound chime is an additional deterrent.

Project Description

The heart of the unit is the chime piezo sounder that is activated by a photocell. At a certain light level, the chime will sound until the light source is reduced below the sensitivity level of the system. The sound output from the chime is a "ding-dong, ding-dong" until light is removed from the photocell. The photocell is sensitive enough, with the aid of a simple lens, to activate the sounder chime from as far away as 200 feet.

Circuit Description

The circuit hook-up for the light sensor alarm is shown in Fig. 9-7.

Basic Light Sensor. It is a rather simple circuit but is powerful in what it does. The circuit is arranged so that when light hits the photocell PC, its resistance drops greatly, from the megohm range to a few tens of ohms. This in effect closes a switch (the low resistance) and applies voltage from the positive side of the battery to the white wire, causing the chime to sound twice. It continues to ring until light is removed from the photocell, or the resistance of the photocell rises to the value that removes the trigger voltage from the white control wire to the sounder.

A push-to-test switch SW is used to test the system without applying light to the photocell. This test also gives you a good indication of the condition of the battery. The system operates with good volume from the sounder using a 6- or 9-volt transistor battery.

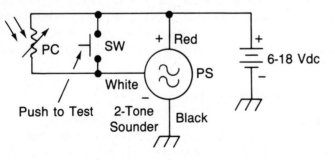

Fig. 9-7. Two-tone sounder rings as long as light strikes the photocell.

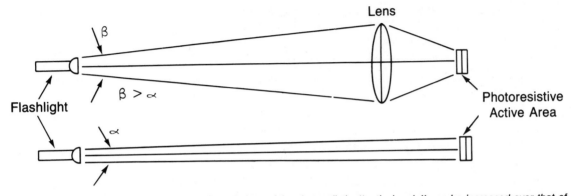

Fig. 9-8. By placing a small magnifying glass (lens) in front of the photocell, the "optical gain" can be increased over that of the cell alone.

Power Consumption. The leads from the photocell to the chime sounder can be hundreds of feet long because very little current is drawn during operation. Data on the 1.5- to 3-volt dc chime (Radio Shack No. 273-067) shows that the total current drawn in the dark (no-sound operation) is just 0.002 mA. In the light, when the sounder is in operation, the total current drawn is less than 0.2 mA.

Chime Control. The chime sounder has a control wire (white) that, when connected to the battery (red) wire for about a tenth of a second, causes the chime to sound twice. It then stops, awaiting another short pulse of voltage to the white control wire. When the white and red wires are connected together, the chime rings continuously.

Magnifying Lens. By placing a small magnifying lens (1 to 2 inches in diameter) in front of the photocell, the optical gain can be increased over that of the photocell alone. Figure 9-8 shows this arrangement using an inexpensive magnifying glass. The glass lens, larger than the photocell, "collects" light and semi-focuses it on the active area of the photoresistor. The light does not have to be in focus on the active area of the photocell. The system arrangement shown in greater detail in Fig. 9-9 sounds when light from a two-cell flashlight strikes the lens from over 200 feet away at night.

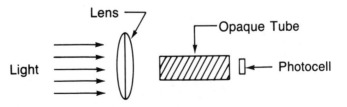

Fig. 9-9. A lens and opaque tube help increase sensitivity of the light sensor alarm.

Table 9-7. Parts List for Light Sensor Alarm.

Item	Description
PC	Photocell, cadmium sulfide (CdS),
	Radio Shack No. 276-116, or equiv.
PS	Piezo sounder, Chime, 6 to 18 Vdc
	Radio Shack No. 273-071 or equiv.
SW	Switch, SPST, momentary push button
Misc	Case, hook-up wire, battery hook-up

A small opaque tube about ½ inch in diameter and 2 to 4 inches long can be placed in front of the photocell to reduce the amount of undesired outside light that might strike the cell at night. This arrangement, shown also in Fig. 9-9, helps to increase the overall light sensitivity of the system.

Parts List

The parts list for the light sensor alarm is shown in Table 9-7. The piezo sounder listed operates on 6 to 18 volts dc and is Radio Shack No. 273-071. The chime sounder operates on 1.5 to 3 volts dc.

Further Uses of Project

The light sensor alarm can be used for additional tasks as shown in Fig. 9-10 where reflected light is used for inspection purposes or for sensing the height of liquid or grain in a bin. Other uses will come to mind as you become familiar with the technique.

MULTIPROBE PIEZO SOUNDER

The multiprobe piezo sounder allows you to use a number of different probes to sense an event and then sounds a "ding-dong" alarm to alert you to the event.

Project Description

The multiprobe contact points are used to trigger a two-transistor circuit that sets off the sounder. The inputs can be two water probes that sound an alarm when a liquid has reached a certain level in a container. This can be used when filling a fish aquarium, a cup of tea for a blind person, or a bathtub. When installed on the floor of a

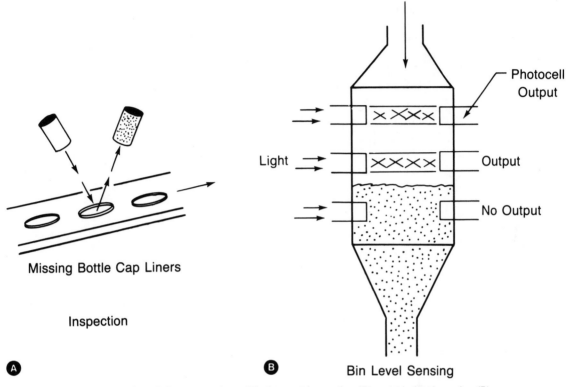

Multiprobe Piezo Sounder

Photocell
Output

Light →

Output

No Output

Missing Bottle Cap Liners

Inspection

Bin Level Sensing

A

B

Fig. 9-10. Additional uses of the light sensor alarm. Missing part inspection (A) and bin level sensing (B).

basement, it can be used to alert you to the fact that water is coming in or that a pipe or connection has broken.

A push button installed on your business counter can be used to alert you to the fact that a customer desires service in case you are in the back room. A momentary push button is used in this case. You can also use a reed relay and magnet to alert you to the fact that a window has been opened. You can use a photocell to let you know when light is present at a particular location. A photocell can be installed in a garage or work area and wires run to the alarm. When a switch is activated that connects the photocell to the circuit, the alarm will sound to remind you to turn the light either on or off. A mercury tilt switch can be mounted on some object to alert you to the fact whether the object is level or not. If someone were to move the object, the mercury switch would turn on and the sounder would activate. Once you have gotten the multiprobe piezo sounder to work for you, you'll think of other uses for it.

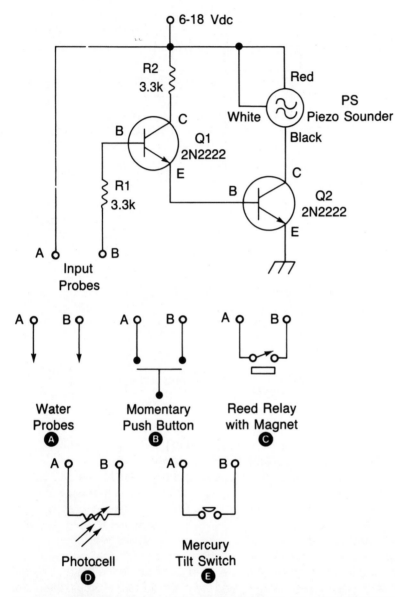

Fig. 9-11. Multiple probe piezo sounder with probe input for use with liquid level (A), momentary push button (B), window or door reed relay (C), photocell input (D), mercury tilt switch (E). Circuit operates when input (A-B) goes toward low resistance.

Circuit Description

The circuit for the multiprobe piezo sounder is shown in Fig. 9-11. It uses two npn transistors (2N2222). The circuit operates in the following

178

manner. When a short (or low resistance) appears at the input probe points A and B, a positive voltage is applied to the base of transistor Q1 so the base to emitter is positively biased. Transistor Q1 conducts, which applies a positive voltage to the base of Q2 so that it also conducts. When the collector of Q2 draws current, it turns on the piezo chime and it begins to sound out.

When a brief contact short is applied across points A and B, the sounder will chime twice and then shut off. If contact is maintained, it will continue to sound out until the contact is removed.

Parts List

The parts list for the multiprobe piezo sounder is shown in Table 9-8. The various sensor contacts are inexpensive and available at most electronics supply stores. Buy and put together those sensors you want to try.

Further Uses of Project

The circuit simply operates when a near short appears across the input probes at points A and B (in Fig. 9-11). You can experiment with other electrical items that produce a change in resistance with a change in input factor such as temperature, humidity, etc.

RESISTANCE-SENSITIVE AUDIO CONTINUITY TESTER

This project describes a continuity tester that changes an audible tone frequency with a change in resistance. The tone goes down in

Table 9-8. Parts List for Multiprobe Piezo Sounder.

Item	Description
PS	Piezo sounder, Chime, Radio Shack No. 273-071 or equivalent
Q1, Q2	Transistor, npn 2N2222
R1, R2	Resistor, 3.3k, ½ watt
Case	Utility case, Radio Shack No. 270-230 or equivalent
Misc	Hook-up wire, solder, PCB, battery holder, reed relay and magnet, push button switch, mercury switch

frequency as the resistance goes up. a short at the test probes produces a high, steady tone. Use this device for testing purposes in the shop.

Project Description

A 555 timer IC is used to produce an audio tone through a small loudspeaker when a resistance is placed across its test probes. Input resistance can vary from 0 ohms, with the test probes shorted, to about 30 megohms. With the test probes shorted, the tone produced is about 7,000 Hz (the condition desired for checking fuses, low-value resistors, inductors, diodes, etc.). When testing high resistance, an upper value of between 20 and 30 megohms produces an intermittent "pop" of about one pulse per second. There is no tone output when the test leads are open.

Circuit Description

The circuit for the resistance-sensitive audio continuity tester is shown in Fig. 9-12. The heart of the circuit is IC1, a 555 timer configured as an astable audio oscillator. The unit is especially handy for testing devices without having to glance from the test probes that are

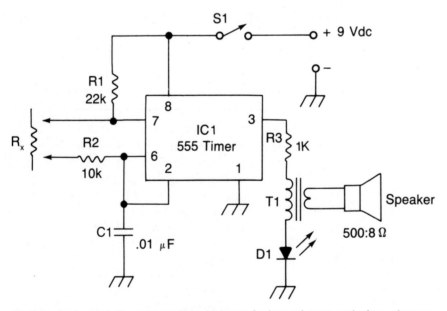

Fig. 9-12. Audio continuity tester produces high tone for low resistance and a few pulses per second for high resistance.

connected to a test point to an ohmmeter and back again. Here you let your ears do the walking as the unit sounds out telling you whether you have a low or high resistance across the test probes.

Figure 9-12 shows where the unknown resistor, R_x, is connected into the oscillator circuit to change its tone according to resistance size. The unknown can have any resistance value from 0 ohms to about 30 megohms. At 0 ohms, the high frequency of 7,000 Hz sounds like a continuous tone to the ear, though it actually is a rapidly changing on/off signal. Only at high resistance values can you hear the intermittent "pop" of a few pulses per second. If you look at the speaker cone when testing these high resistances, you can see it draw in and out with each pulse.

Very little current flows through the test probes when in use. When R_x is 0 ohms, the current level through the probes is about 270 microamperes, and when R_x is 1 megohm, the current is about 9 microamperes.

The 555 timer is operated in the astable oscillator mode. The free-running frequency and duty cycle are both accurately controlled with three resistors and one capacitor. The external capacitor, C1, charges through R1, R2, and R_x, but it discharges only through R2 and R_x. R2 limits the upper frequency of oscillation to about 7,000 pulses per second to keep the tone range within that of human hearing. The lower frequency limit of approximately one pulse per second is set by the value of R_x when above about 20 megohms.

Resistor R3 limits the current drawn through the output circuit. A value of 1 kilohm provides adequate audio volume with a 9-volt battery. The LED, D1, acts as a pilot light and flashes on and off at low frequencies about 10 to 12 pulses per second. Otherwise, it appears to be on all the time.

Parts List

The parts list is shown in Table 9-9. There are no unusual parts required.

HEADLIGHT REMINDER

This project is a simple reminder to turn on your headlights when driving in your car as dusk approaches.

Project Description

The headlight reminder is most important during the early evening hours as you are driving in your car and dusk approaches. This handy

Table 9-9. Parts List for Resistance-Sensitive Audio Tester.

Item	Description
IC1	555 timer IC
C1	Capacitor, 0.01µF, 35 Vdc
D1	Green or red LED
R1	Resistor, 2.2k, ½ watt
R2	Resistor, 10k, ½ watt
R3	Resistor, 1k, ½ watt
S1	Switch, SPST, on/off
Misc	Plastic case, battery connector, hook-up wire, solder, 8-pin IC Chip, PCB, test leads

project turns on a sounder as it begins to get dark to remind you to turn on your headlights. At dusk, people often drive around with their headlights off, not knowing how difficult it is to see them. This headlight reminder will prevent you from making the same mistake.

The unit has a sensitivity control that you adjust initially as dusk approaches. When it gets dark enough for the setting of the control, the sounder comes on.

Circuit Description

Figure 9-13 shows the schematic circuit for the headlight reminder. In the daytime, the circuit is energized when the ignition switch is turned on. However, the photocell PC is in daylight and has a low resistance across its terminals. As dusk approaches, the resistance of the photocell begins to rise so that a greater positive voltage appears at the top of the photocell, which is connected to the gate of the silicon-controlled rectifier (SCR).

As it continues to get darker, there is a time when the photocell resistance has risen to a value where there is a larger voltage drop across its terminals. Through voltage divider action across the photocell, the gate of the SCR becomes positive enough and the SCR triggers on. The sounder is thus turned on and chimes until you turn on the headlight switch. With the headlights on, points A and B both have 12 volts dc and the circuit is deactivated.

The sensitivity of the circuit is determined by the setting of resistor R1, the 100k potentiometer. It might take a few tries (evenings) to get

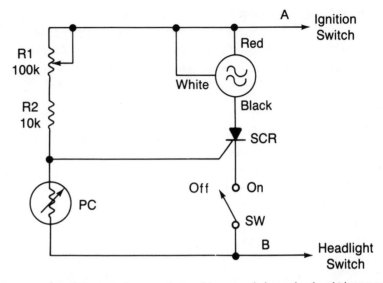

Fig. 9-13. Headlight reminder uses photocell to sense darkness level and triggers on piezo sounder. Sensitivity is set by R1.

this control set to the proper level so the sounder comes on before it gets too dark.

Parts List

The parts list for the headlight reminder is shown in Table 9-10. The components can be mounted in a small plastic box and mounted on the top of the dash where you can easily reach the sensitivity control to adjust the sensitivity level.

Table 9-10. Parts List for Headlight Reminder.

Item	Description
PS	Piezo sounder, Chime, Radio Shack No. 273-071 or equivalent
R1	Resistor, Potentiometer, 100k
R2	Resistor, 10k, ½ watt
SCR	Silicon-controlled rectifier, Radio Shack No. 276-1162 or equiv.
SW	Switch, on/off
Misc	Case, hook-up wire, solder

Further Uses of Project

In addition to reminding you to turn on the headlights in your car or truck, you can also use the project in your home or business. Use it with a 9-volt battery installed to remind you to turn on the outside lights, bring in the cows or chickens, or to secure any outside perimeter protection.

LAMP DIMMER AND MOTOR CONTROL

With this easy-to-construct project, you can turn on and adjust the light level of an incandescent lamp from your easy chair. You can also control the speed of a small ac motor.

Project Description

Incandescent lamps sometimes have a habit of burning out at the most inconvenient times. This lamp dimmer (and motor control project) is one way of reducing the possibility of this problem. The heart of the project is a triac, the ac version of the SCR. The triac is turned on each half cycle of the ac voltage by a diac, which is a small solid-state device resembling two ordinary back-to-back diodes.

The unit can be packaged in a small plastic case and is placed handily on a stand by your easy chair. With a potentiometer control, you can adjust the light level of a table lamp plugged into the unit from full off to full on. This gives you a full span of control of the light level from your lamp as well as extending the life of the bulb, because you can always turn it up from a lower wattage level to the full brilliance level. You then spare your lamp filament from "thermal shock" and it will have a much longer life span.

The lamp dimmer is for incandescent bulbs only, not fluorescent lamps. Additionally, the motor used with the lamp dimmer must be of the brush-type such as that found in the more common hand tools, kitchen appliances, sewing machines, saber saws and the like.

Circuit Description

The circuit diagram for the lamp dimmer and motor control is shown in Fig. 9-14. The center of the circuit is the triac TR that is turned on by the diac D. The circuit works in the following manner. The triac is off when there is no voltage on gate G. It turns on when a voltage pulse of sufficient amplitude is applied to the gate through the diac. The diac is a bilateral trigger diode. The timing of this pulse is established by the time constant of resistor R1 and capacitor C. When the voltage pulse is large enough, it causes the diac to conduct in both

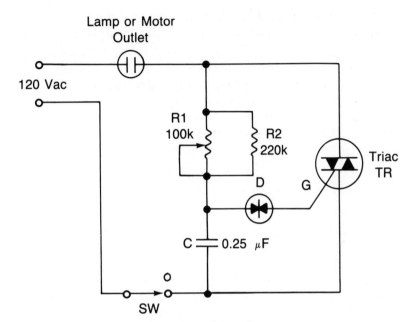

Fig. 9-14. Lamp dimmer and motor speed control.

directions and turns on the gate. The triac is triggered on by the gate voltage and stays on for a percentage of time, depending on the RC time constant. The triac acts like an ac switch, turning on for part of the time during each half of the ac cycle.

The lamp or motor to be controlled is plugged into an outlet, and you can control up to about 300 watts incandescent, or a ¼ horsepower motor such as a drill, small saber saw, or soldering iron.

Wire the ac part of the circuit with care and do not leave any exposed wires that might shock you. Wrap a layer of electrical tape across or around any exposed wires that carry the power to and from the triac.

Parts List

The parts list for the lamp dimmer is shown in Table 9-11. All parts are readily available at local electronics stores.

ULTRASONIC TRANSMITTER AND RECEIVER

This project describes an ultrasonic transmitter and receiver that can silently turn on a relay or activate a sounder chime by pressing a button. The project has applications in the home, security devices, and experimentation.

185

Table 9-11. Parts List for Lamp Dimmer.

Item	Description
C	Capacitor, 0.25μF, 600 Vdc
D	Diac, bilateral trigger diode
O	Outlet, 120 Vac
R1	Potentiometer, 100k
R2	Resistor, 220k, ½ watt
SW	Switch, on/off, 120 Vac, 5A
TR	Triac, Radio Shack No. 276-1000
Misc	Case, hook-up wire, hardware, line cord, solder

Project Description

This project lets you try your hand at working with ultrasonic sound waves. Two electronic packages are to be built. One is an ultrasonic transmitter that puts out a 40-kHz sound wave from a piezo electric sounder when you press a button. The second package is an ultrasonic receiver that uses an identical sounder as the transmitter. However, the sounder, or transducer, acts as a microphone tuned to the 40-kHz signal that the transmitter sends out. It amplifies this signal sufficiently to turn on a relay or activate the sounder when you press the button. The relay terminals can turn on a light, a motor, open a lock, or other creative functions.

Circuit Description

Take a closer look at the circuits that go into making up the ultrasonic transmitter and receiver.

Transmitter. The circuit diagram for the transmitter is shown in Fig. 9-15. The transmitter is rather simple compared to the receiver. The main unit is the piezo electric transducer UTR that puts out an acoustic (sound) wave signal at 40 kHz when energized by transistor Q1 and when the push button PB is pushed. Q1 is a 2N2222 npn transistor and it causes the transducer to emit an ultrasonic signal (above normal human hearing range).

A signal will be emitted from the UTR only when the push button is being pressed, so not much power is consumed from the 9-volt battery. The sounder is activated at the receiver each time the transmitter push button is pressed.

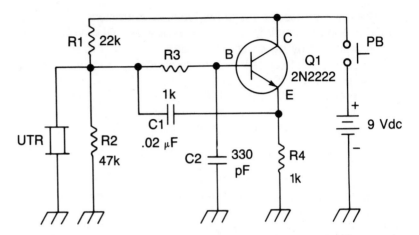

Fig. 9-15. Ultrasonic transmitter can activate remote receiver relay or a piezo sounder.

The transmitter can be easily mounted in a small plastic case so that it can be hand-carried to different locations that are within range of the receiver.

Receiver. The receiver circuit diagram is shown in Fig. 9-16. The heart of the receiver is the ultrasonic transducer UTR, which is identical to that used in the transmitter. However in the receiver, the UTR is used as a receiving element, or microphone, for the 40-kHz signal.

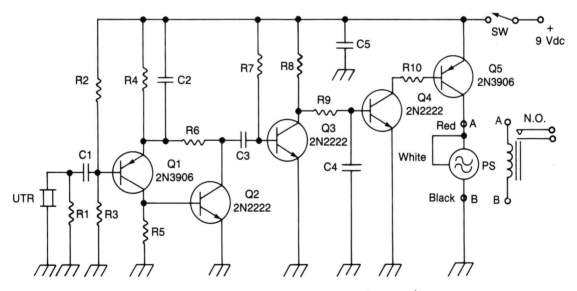

Fig. 9-16. Ultrasonic receiver at 40 kHz can activate piezo sounder or normally open relay.

The receiver uses two different types of transistors, a pnp (the 2N3906), and an npn (the 2N2222). The weak signal received by the UTR is amplified by transistors Q1 through Q5 until the signal is of sufficient amplitude to close the normally open relay or to activate the piezo sounder PS.

The receiver should be turned off until you plan to use it for a period of time. The receiver must be on all the time for it to be active and ready to receive a signal from the transmitter.

Parts List

The parts list for the transmitter is shown in Table 9-12 and the parts list for the receiver is shown in Table 9-13. Parts are readily available at local electronics supply stores. The 40-kHz transducer is available from DC Electronics as Ultrasonic Transducer, Catalog No. J4-815, DC Electronics, P.O. Box 3203, Scottsdale, AZ 85257. The cost is approximately $10.00. The DC Electronics toll-free number is 1-800-423-0070.

Further Uses of Project

The project is especially good for remote switching or control of radios, TVs, alarms, etc. Because the transmitted signal is like a sound wave except ultrasonic, there must be a good air path from the transmitter to the receiver. That is, the signal will not be able to pass through a closed door or window. Depending on the distance, the signal should be able to wind its way around bends and turns in a hallway or upstairs. A bit of experimenting will tell you what distances you can cover and still maintain good signaling.

Table 9-12. Parts List for Ultrasonic Transmitter.

Item	Description
C1	Capacitor, 0.02 μF, 35 Vdc
C2	Capacitor, 330 pF, 35 Vdc
Q1	Transistor, npn, 2N2222
R1	Resistor, 22k, ½ watt
R2	Resistor, 47k, ½ watt
R3, R4	Resistor, 1k, ½ watt
PB	Push button, momentary switch
UTR	Ultrasonic transducer, 40 kHz
Misc	Case, hook-up wire, solder, hardware

Table 9-13. Parts List for Ultrasonic Receiver.

Item	Description
C1	Capacitor, 0.01 μF, 35 Vdc
C2	Capacitor, 10 μF, 35 Vdc
C3	Capacitor, 0.01 μF, 35 Vdc
C4, C5	Capacitor, 100 μF, 35 Vdc
PS	Piezo sounder, Chime, Radio Shack No. 273-071, 6 to 18 Vdc
Q1, Q5	Transistor, pnp, 2N3906
Q2, Q3, Q4	Transistor, npn, 2N2222
R1	Resistor, 2.2M, ½ watt
R2	Resistor, 22k, ½ watt
R3, R8	Resistor, 47k, ½ watt
R4, R6	Resistor, 1k, ½ watt
R5, R9	Resistor, 3.3k, ½ watt
R7	Resistor, 150k, ½ watt
R10	Resistor, 330, ½ watt
RY	Relay, 6 Vdc, normally open
SW	Switch, on/off, SPST
UTR	Ultrasonic transducer, 40 kHz
Misc	Cabinet, 9-volt power supply, plugs, jacks, hook-up wire, hardware, solder

NIGHTLIGHT TURN-ON

The nightlight turn-on is an easy-to-build project that turns on a small nightlight when a room darkens in the evening. The unit consumes little power and is left plugged into a wall outlet.

Project Description

The simple circuit used in this nightlight project will help you understand how these small lights operate whenever a photocell is darkened by a passing shadow in the daytime or at night when the room darkens. A small 7½-watt bulb is used and provides enough light for use in hallways or as a nightlight in a child's room.

Circuit Description

The circuit diagram for the nightlight turn-on is shown in Fig. 9-17. Notice it uses an npn transistor to turn on the small lamp rather than

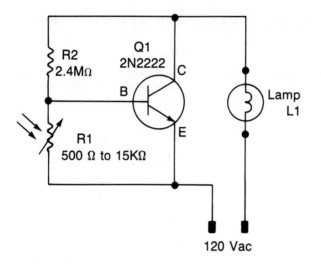

Fig. 9-17. Nightlight with photocell turn-on.

a triac as the larger wattage circuits do. The circuit operates in the following manner. In the daylight, there is much light on the photocell (photoresistor), R1, and its resistance is at a very low value (approximately 500 ohms). This means that the base of the transistor Q1 (a 2N2222) is almost at the same potential as the emitter. The transistor is therefore cut off and there is no current through the transistor. The lamp, therefore, remains off.

As the light level on the photoresistor R1 begins to decrease, its resistance begins to increase. Through voltage divider action between R1 and R2, a larger positive voltage drop begins to appear from the base to the emitter. As the base becomes more positive with respect to the emitter, it approaches the point where the transistor begins to conduct, when the base-to-emitter voltage approaches about 0.7 volts dc, with the base positive. In total darkness, the value of the photoresistor reaches a maximum of 15 kilohms, and through voltage divider action, the base is as positive as it can become with respect to the emitter. The transistor is driven into conduction and collector current flows through the lamp L1 to light it to maximum brilliance.

The transistor is on and conducts only when the line voltage is of such a polarity that the base is positive with respect to the emitter. There is no time delay in the circuit, and the lamp can be turned on and off by blocking the light to the photoresistor. Do this by waving your hand in front of the photocell to block out light, or you can shine a flashlight on the cell and the lamp will turn off.

Table 9-14. Parts List for Nightlight with Photocell.

Item	Description
L1	Lamp, 7½ watt, 120 Vdc
Q1	Transistor, npn, 2N2222
R1	Photocell, Radio Shack No. 276-116, or equivalent
R2	Resistor, 2.4M, ½ watt
Misc	Case, ac cord, hook-up wire, lamp socket

Parts List

The parts list for the nightlight turn-on is shown in Table 9-14.

LOW-LEVEL AUDIO AMPLIFIER

This small audio amplifier is for use around the house or workshop, including automotive work. When using the inductive pickup, you can listen to the electrical system of an auto or truck to determine how it is operating. You will find that you use the amplifier frequently around the shop for experimenting with other projects.

Project Description

A small, hand-held audio amplifier is a handy thing to have around the workbench while working on different projects. This small amplifier provides enough gain to use with various input transducers such as a small microphone, a telephone inductive pickup, phonograph crystal pickup, a solar cell to "listen" to light, or a hydrophone to listen to fish in your aquarium.

The audio output can be a small loudspeaker, or you can use a small transistor radio earpiece. You can also use this small amplifier to feed a larger amplifier or public address system. It is battery-powered, so it can be mounted in a small plastic case so that you can take it with you.

Circuit Description

The circuit for the small audio amplifier is shown in Fig. 9-18. It uses an npn 2N2222 transistor for one stage of amplification. The audio

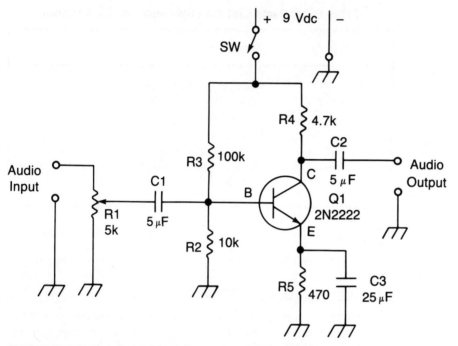

Fig. 9-18. Low-level audio amplifier using common emitter transistor.

input is fed to the base of the 2N2222 through capacitor C1 and the audio output is taken from the collector through capacitor C2. Operating bias for the base is obtained from the center point of resistors R2 and R3 through voltage divider action.

Parts List

The parts list for the audio amplifier is shown in Table 9-15 and all the parts are standard. The complete circuit will mount easily in a small plastic case so that it can be hand-carried.

Further Uses of Project

This project can help you learn about audio amplifiers and how they operate. With a telephone inductive pickup as input to the amplifier, you can use the two to look for the location of house wiring behind walls by listening for hum caused by the electrical wiring. You can locate a wire within ¼ of an inch and tell on which side of the wall stud the wiring runs. This is important when hanging curtains or doing any housework where you might drill or nail into a stud that has wires run-

Table 9-15. Parts List for Low-Level Audio Amplifier.

Item	Description
C1, C2	Capacitor, 5 µF, 35 Vdc
C3	Capacitor, 25 µF, 35 Vdc
Q1	Transistor, npn, 2N2222
R1	Resistor, Potentiometer, 5k
R2	Resistor, 10k, ½ watt
R3	Resistor, 100k, ½ watt
R4	Resistor, 4.7k, ½ watt
R5	Resistor, 470, ½ watt
SW	Switch, on/off, SPST
Misc.	Cabinet, hook-up wire, battery connector, input and output jacks

ning up the side. A good idea is to use an earphone to cut out outside noise.

BATTERY ELIMINATOR

The battery eliminator is a transformer power supply that provides 9 volts dc output for use with transistor-equipped devices such as transistor radios and tape recorders. The device operates on 120 Vac house voltage.

Project Description

This easy-to-build power supply will help you save on battery costs because you can use it with any battery-operated device that requires 9 volts. This includes small CB radios, pocket calculators, stereo receivers, and any device that does not draw more than 300 mA from house current. The power supply is plugged into any 120 Vac outlet. Hence, you have continuous power, whereas with batteries, their condition deteriorates whether they are used or not.

Circuit Description

The battery eliminator schematic is shown in Fig. 9-19. The 120-volt house current is stepped down by the transformer T to about 6.3 Vac. The transformer isolates the line voltage from the supply, which makes the supply safe to use. The 6.3 Vac is rectified by the bridge rectifier

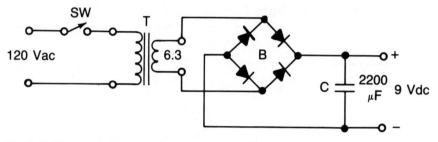

Fig. 9-19. Battery eliminator will deliver 9 Vdc from 120 Vac.

B, which consists of four diodes in one plastic package. Note that the positive terminal of the supply is taken from the two cathodes of the output diodes in the bridge.

The circuit, in connection with the large capacitor C provides smooth direct current with minimum ripple. Therefore, there is a minimum hum when used with audio devices. When connecting the ac line cord to the primary connection leads, solder them and then wrap with electrical tape to prevent electrical shock or a short with other components.

The dc connection to the device to be powered can be made through a mini, sub-mini, or power supply plugs. If your set has no power jack, you can wire the circuit into the battery connections directly.

Parts List

The parts list for the battery eliminator is shown in Table 9-16. Parts are readily available at most electronics supply stores.

Table 9-16. Parts List for Battery Eliminator.

Item	Description
B	Rectifier, bridge, full-wave, Radio Shack No. 276-1171 or equiv.
C	Capacitor, 2200 μF, 50 Vdc, Radio Shack No. 272-1048, or equiv.
SW	Switch, on/off, SPST, 120 Vac
T	Transformer, 120 Vac to 6.3 Vac
Misc	Case, hook-up wire, hardware, line cord, solder

Further Uses of Project

The primary use of the power supply is to eliminate batteries that may be required by the device you are using. However, it can be used in many of the circuits described so far in this chapter if you want them to operate directly from the house current.

TELEPHONE ON-HOLD PROJECT

The telephone on-hold project lets you place a phone line on hold while you walk to another room to answer from another phone.

Project Description

Say the phone rings, you answer it, and it is for someone else in the house or office. If you don't have the capability of placing the line on hold, you have to put the instrument down, find the called person and then go back to hang up the phone after the called person has picked up the phone. There is a lot of walking and talking involved in "transferring" a call from one phone to another.

With the telephone on-hold project installed you answer the phone and find out that you want to put the line on hold. You momentarily simply press a push button while you hang up the phone. This places the line on hold and also lights an LED to indicate such. You can then pick up from any phone on that line.

Circuit Description

The circuit for the phone on-hold project is shown in Fig. 9-20. The center of the circuit is the SCR, whose action places a 1200-ohm resistor across the telephone line to place it on hold. The circuit operates as follows. When the phone is on the hook, the voltage across the line is 48 Vdc. No current flows through the circuit, because the SCR is off (open). When the phone is off the hook and in use, the voltage across the line is about 6 Vdc. There is still no current flow through the circuit, because the SCR is still off.

When you want to place the line on hold, press the momentary push button PB and continue to hold it as you hang up. When the switch is pressed, R1 and R2 momentarily form a voltage divider that applies a portion of the voltage across the telephone line to the gate of the SCR. With the gate positive, the SCR is triggered on. When the SCR triggers, the 1200-ohm resistor R1 is placed across the line, which puts it on hold and prevents a disconnect. The calling party can hear nothing when on hold.

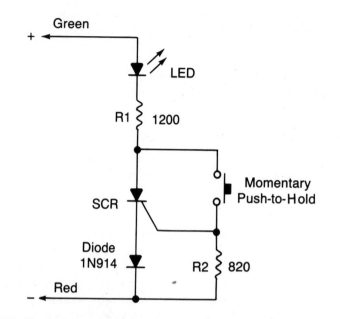

Fig. 9-20. Circuit to place telephone line on hold. It can be answered from any phone on the line.

The LED in series with the 1200 ohm resistor now has current flowing through it and it is now lighted to indicate the line is on hold. The LED will stay lighted as long as the line is on hold to serve as a reminder.

When any instrument on the line is picked up, the line voltage drops to the nominal 6 Vdc, the voltage across the SCR decreases, the SCR releases the 1200-ohm resistor from the line, and the LED then goes out. The circuit is now ready for another call and possible use. The diode, 1N914, serves to protect the circuit from reverse voltage that might appear across the line. Correct line polarity must be observed when making the installation to your telephone line.

Parts List

The parts list for the telephone on-hold is shown in Table 9-17. All parts are readily available at local electronics supply stores.

Further Uses of Project

There are no further uses of the project because it is specialized for use across a telephone line. However, as you gain experience about

Table 9-17. Parts List for Telephone Line Hold.

Item	Description
LED	Green or red LED
Diode	Diode, 1N914 or equivalent
R1	Resistor, 1200, ½ watt
R2	Resistor, 820, ½ watt
SCR	Silicon-controlled rectifier, 200 V, 6 A, Radio Shack No. 276-1067 or equivalent
Misc	Case, modular telephone plug, hook-up wire, solder

the circuit and telephone electronics, you will be able to devise other applications for parts of the circuit, as in the project that follows.

VISUAL TELEPHONE RINGER

The visual telephone ringer lights a lamp instead of ringing a bell or buzzer. This is handy when you want quiet in a house or office or, conversely, when excess noise would overpower a normal ringing telephone.

Project Description

This project is connected to the telephone line, to provide the calling signal, and to the house current, to provide voltage to turn on a 120-volt ac lamp. When a call comes in, the lamp lights each time it "rings," and the flashing light will catch your attention.

Circuit Description

The circuit diagram for the visual telephone ringer is shown in Fig. 9-21. An optical coupler is used to connect to the telephone line to signal when the line is ringing. This in turn switches on the triac to supply 120 Vac to the incandescent lamp to turn it on. The light switches on only when the phone rings—on for 2 seconds, off for 4 seconds—until the phone is answered.

As shown in the figure, the light ringer is connected to the telephone line by means of a modular plug. This connection can be

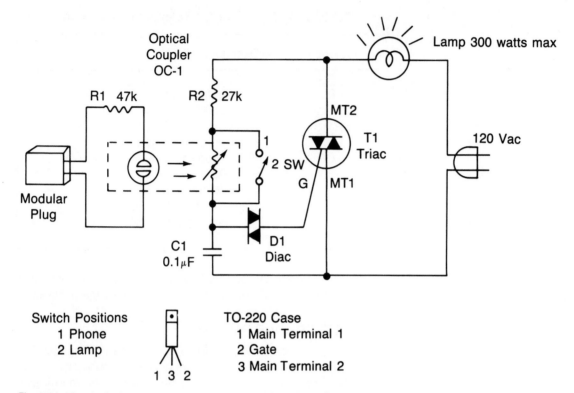

Optical
Coupler
OC-1

Lamp 300 watts max

R1 47k

R2 27k

MT2

T1
Triac

120 Vac

Modular
Plug

1

2 SW

G MT1

C1
0.1μF

D1
Diac

Switch Positions
1 Phone
2 Lamp

TO-220 Case
1 Main Terminal 1
2 Gate
3 Main Terminal 2

1 3 2

Fig. 9-21. Visual telephone "ringer" lamp lights each time phone rings.

made anywhere you want the lamp to light. The resistor R1 and the neon bulb are connected across the line. With the phone on the hook, no current flows in the circuit because there is only 48 Vdc on the line. It takes about 65 volts to ionize the neon bulb, so it remains off.

When the signal voltage comes in, it swings plus and minus about 90 volts, so the neon bulb will light. When the bulb lights, the resistance of the closely coupled photoresistor drops greatly and voltage appears across the diac D1. This causes the gate G of the triac to go positive and the triac turns on. When the triac turns on, the resistance from terminals MT1 and MT2 of the triac drops to near zero. This action completes the circuit to the incandescent lamp L1, and 120 Vac is applied to the lamp, causing it to light to full brilliance while the phone is ringing.

The triac swings on for each half cycle (plus and minus) of the 60-Hz power line voltage because the gate is still positive as controlled by the diac and photoresistor/neon bulb combination. Remember, the

Table 9-18. Parts List for Visual Telephone Ringer.

Item	Description
C1	Capacitor, 0.1 μF, 100 Vdc
D1	Diac, 1N5758 or equivalent
NE-2	Neon bulb, NE-2
OC-1	Optical coupler, Sigma 301T1-120A1 (if not available, See NE-2 and R3)
R1	Resistor, 47k, ½ watt
R2	Resistor, 27k, ½ watt
R3	Resistor, Photocell, Cadmium-Sulfide (CdS), Radio Shack No. 276-116 or equiv.
T1	Triac, 400 Vdc, 6 A, Radio Shack No. 276-1000, or equivalent
Misc	Modular telephone plug, case, hook-up wire, ac receptacle, ac line cord with plug

ringing voltage is 90 Vrms at a 20 Hz rate, while the power line voltage is 60 Hz—three times as fast.

The lamp to be controlled is plugged into an ac outlet and is available at most hardware or general appliance stores. The lamp should be switched on and left on, because power is applied only when the phone "rings." Be sure to insulate the components, and be aware of exposed points where 120 Vac is present when installing the project in a metal or plastic utility case or cabinet. An ac cord connects the unit to any ac outlet to power the lamp.

Parts List

The parts list for the visual telephone ringer is shown in Table 9-18. All parts are readily available at electronics stores.

Further Uses of Project

The project can be used to turn on any device that operates on 120 Vac when the phone rings. However, the circuit does not provide a latching action so that power would remain on after the phone rings. Total power available is about 300 watts at the ac outlet jack, as this is the power rating for the triac.

INDEX

INDEX

200-in-1 electronics project kit (Radio Shack), 135
60-project electronic lab (Radio Shack), 135
75-meter Compact SSB transceiver (Radiokit), 133

A

ac-dc Fundamentals (EKI), 132
acceptor elements (doping), 21
  electrons in shell of, 22
active components, 24
adjustable power supply (Jameco), 138
Advanced Digital and Operational Amps (EKI), 133
air inductor, 47
Ampere, Andre Marie, 25
Amplifiers and Oscillators Course (EKI), 132
Analog Electronics course (EKI), 133
atomic structure, 13
  conductors in, 18
  copper, 15, 17
  electrons in, 16
  elements and, 14
  helium, 14
  hydrogen, 13
  insulator in, 19
  semiconductors in, 19
attraction, law of, 14

audio amplifier (Jameco), 141
audio amplifier/intercom (EKI), 131
audio amplifiers, symbols for, 51
audio continuity and voltage tester, 155-157
audio oscillator (Radiokit), 134
audion, 33
auto/home burglar alarm (Jameco), 140
automatic siren (EKI), 129

B

Bardeen, John, 34
barrier blocks, 58, 59
base, 35
battery checker (Jameco), 141
battery eliminator, 193-195
big sound portable organ (EKI), 131
binary clock (EKI), 132
bipolar transistor, 35
block diagrams, 41
body blinker (EKI), 132
braid, desoldering, 73
Brattain, Walter, 34
breadboarding, 89-93
burglar alarm (EKI), 128

C

capacitance, 28
capacitance substitution kit (Jameco), 140

capacitors, 28, 38
  charge in, 29
  dielectric, 29
  schematic symbols for, 46
car alarm MK2 (DSE), 135
car horn, musical (Jameco), 143
catswhiskers, 34
CFM decoder (Jameco), 141
charge, electrical, 13
Christmas tree (EKI), 129
CIR-KITs, 85
circuit components, symbols for, 52
circuit connections, symbols for, 51
circuit design, 106
circuit diagram, SCR switch, 110
cleaning up PCBs, 126
clock kit (Jameco), 136
collector, 35
color code
  resistor, 26, 27, 28
  wire, 87
color organ (EKI), 130, 131
color organ (Jameco), 139, 142
combination lock/alarm control (EKI), 130
commercial assembly kits (see kits), 127-151
complex sound generator (Jameco), 141
components
  mounting of, 101-105, 124

proper handling of, 8
soldering of, 124
symbols for, 39
conductors, metal, 18
continuity/polarity tester and LED
    power supply, 157-163
copper atom, 15, 17
copper wire, specifications for, 88
crystal detector, 34
crystal diodes, 34
current, 16, 17, 25, 31
current sources, symbols for, 50
custom-etched PCBs, 101

D

DASH-PCB, 122
dc resistance, 30
De Forest, Lee, 33
decision maker (EKI), 128, 129
decision maker (Jameco), 139
design, 2
safety in, 11
desoldering, 72-78
    braid for, 73
    heat and pull method in, 72
    heat and tap method in, 73
    hot tweezer for, 75
    ICs, tip for, 73
    iron for, 76
    solder sucker in, 73
    station for, 76
    techniques for, 77
detinning, 79
Dick Smith Electronics, 135
dielectric, 29
diffusion, 36
digital bird (EKI), 131
digital clock (EKI), 131
digital dice (EKI), 131
digital dice (Jameco), 142
Digital Electronics and IC course
    (EKI), 133
digital key (Jameco), 141
digital roulette (EKI), 130
digital slot machine (EKI), 131
digital thermometer (Jameco), 137
digital trainer (EKI), 132
diodes
    crystal, 34
    semiconductor, 33
    symbols for, 50
direct-etch PCB for simple circuits,
    119-121
    cleanup of, 120
    etching of, 120
    preparing board for, 120
discrete components, 24

layout of, 106
donor elements (doping), 21
    electrons in shell of, 21
doped crystal semiconductor, 21
    acceptor elements for, 21
    donor elements for, 21
double decision maker (EKI), 129
double fuzz box (Jameco), 141
drilling PCBs, 123

E

Edison effect, 31
Edison, Thomas A., 31
electrical properties, 25
electrical units and symbols, 38-40
    designation for multiple
        components, 39
    multiplying factors and prefixes in,
        39
electron theory, 13-22
electron tubes, 31
    Edison effect and, 31
    Fleming valve as, 32
electronic components, 23-40
    discrete, 24
    hybrid circuits as, 24
    passive and active, 24
    types of, 23-24
electronic construction, general
    principles of, 1-12
electronic diagrams, 41-55
    block, 41
    layout, 43
    pictorial, 42
    schematic, 45
    wiring, 44
electronic doorbell (Jameco), 142
electronic project kits (see kits), 3
electronic roulette (Jameco), 141
electronic security photobeam alarm
    (Heathkit/Zenith), 148
electronic tennis (EKI), 130
electronic timer (Jameco), 140
Electronics Kits International, 127-133
electrons
    bumping, 16
    ring or shell of, 17
    structure and placement of, 16
elements, 14
emitter, 35
epitaxial growth, 36
etching, 36, 114, 116
    kits, 119
    TEC-200 Image Film for, 121

F

farad, 28

fish caller (EKI), 129
fish caller (Heathkit/Zenith), 144
fish caller (Jameco), 139
fixed resistors, 45
Fleming valve, 32
Fleming, John A., 32
flood alarm (Heathkit/Zenith), 148
flux, 71
FM wireless mike (Jameco), 141
freezer alarm (Heathkit/Zenith), 148
full-wave motor speed control (EKI),
    131
function generator (Jameco), 137,
    143
fusible resistors, 46
fuzz box (EKI), 129
    double (Jameco), 141

G

GC reversing film, 118
general principles, 1-12
general-purpose preamplifier (DSE),
    136
germanium, 19, 34
Graymark Komponent Kits, 138-144

H

happy face LED blinker (Jameco),
    142
hardware, 56
headlight delay (Jameco), 142
headlight reminder, 181-184
Heathkit/Zenith kits, 144
helium atom, 14
Henry, Joseph, 31
Hewlett-Packard Company, 3
hot tweezer, 75
hybrid circuits, 24
hydrogen atom, 13

I

ignition killer (DSE), 135
impedance, 48
inductors, 30, 38
    air, 47
    henry as unit for, 31
    magentic-core, 47
    schematic symbol for, 47
infrared control link (Radiokit), 134
insulators, 19
    electrons in shells of, 19
integrated circuits, 35-38
    components in, 37
    desoldering of, 73
    special tools for, 66
introduction to electronic components
    (EKI), 132

intrusion alarm trip circuit, 163-165
iron-core transformers, 48
isolators, optically-coupled, symbols
  for, 51

J

jacks, symbols for, 50
junction transistor, 35

K

Kilby, Jack S., 24
kits, 3, 116-119, 127-151
  developing of, 118
  Dick Smith Electronics, 135
  etching of, 119
  exposing P.T.S. film in, 118
  Graymark Komponent, 138-144
  Heathkit/Zenith, 144
  Jameco Electronics, 136
  P.T.S. positive tranfer system for,
    117
  PCS Lab Kit for, 117
  Radio Shack, 134
  Radio Shack PCB, 119
  Radiokit, 133
  transparency application in, 119
knives, 7, 60

L

lamp dimmer and motor control,
  185-185
large-scale integration chips, 38
layout
  diagrams for, 43
  discrete component, 106
  printed circut board, 106
LED flasher (Jameco), 143
LED pendulum metronome (EKI), 128
librarian tormentor (EKI), 132
light meter (Heathkit/Zenith), 149
light sensor alarm, 173-176
line tracer robot (Jameco), 143
linear IC amplifier circuit, schematic
  for, 54
logic probe (EKI), 128
logic probe (Jameco), 141
low-level audio amplifier, 191-193
lugs, 56, 57

M

magnetic fields, 30
magnetic-core inductor, 47
magnetism, 14
masking, 36
medium-scale integration chips, 38
metal conductors, 18
  electrons in shell of, 18

metal detector (EKI), 128
metal detector (Jameco), 139
metronome (EKI), 128
micro soldering iron, 60
Microprocessor and Machine
  Language Programming course
  (EKI), 133
microprocessor course (EKI), 132
mini-wink (EKI), 129
Morse code oscillator (Heathkit/Ze-
  nith), 145
motor speed control (Jameco), 140
mounting components, 124
multiprobe piezo sounder, 176-179
musical horn (EKI), 131

N

n-type material, 21
neon flasher (Jameco), 139
nerve tester (EKI), 131
noise-activated switch and piezo
  sounder, 166-169
nucleus, 15

O

ohm, 26
Ohm, Georg Simon, 26
one-channel color organ (EKI), 129
One-IC radio (Jameco), 141
OSCAR PSK demodulator (Radiokit),
  133

P

p-type material, 22
P.T.S. positive transfer system, 117
passive components, 24
PCS Lab Kit, 117
pentavalent atoms, 21
perforated boards, 93-98
phasor gun/sound generator (EKI),
  132
photobeam alarm (Heathkit/Zenith),
  148
pictorial diagrams, 42
pliers, 7, 60
plugs, symbols for, 50
pocket dice (EKI), 128
point-contact transistor, 34
point-to-point wiring, 89
positive transfer, P.T.S. system for,
  117
positive transparency, 111
  kits, 119
power supply
  regulated (EKI), 129, 130, 131
  variable (Jameco), 140
power supply (Heathkit/Zenith), 151

power supply (Jameco), 139, 141
pre-etched PCBs, 100
preamplifier (DSE), 136
printed circuit boards, 98-101
  clean-up and testing of, 126
  cleaning of, 114
  coating of, 115
  commercial kits for, 116
  course in (Heathkit/Zenith), 144
  custom-etched, 101
  DASH-PCB, 122
  developing of, 115
  direct-etch, 119-121
  drilling of, 123
  drying developed, 116
  drying of, 115
  etching of, 114, 116
  layout of, 106
  manufacturing process for,
    111-116
  mounting and soldering
    components to, 124
  positive transparency in, 111
  pre-etched, 100
  proper handling of, 8
  removing resist on, 116
  repairing damaged or missing
    tracks on, 82
  review circuit for, 111
  TEC-200 Image Film for, 121
  transferring schematic to, 107-110
  types of, 86
  ultraviolet exposure of, 115
printed wiring boards (see printed
  circuit boards)
programmable robot (Jameco), 143
projects, 152-199
  audio continuity and voltage tester,
    155-157
  battery eliminator, 193-195
  continuity/polarity tester and LED
    power supply, 157-163
  fields of application for, 152
  headlight reminder, 181-184
  intrusion alarm trip circuit, 163-165
  lamp dimmer and motor control,
    184-185
  light sensor alarm, 173-176
  low-level audio amplifier, 191-193
  multiprobe piezo sounder, 176-179
  noise-activated switch and piezo
    sounder, 166-169
  resistance-sensitive audio continuity
    tester, 179-181
  telephone on-hold, 195-197
  ultrasonic transmitter and receiver,
    185-191

visual telephone ringer, 197-199
whistler lighting receiver, 169-173
proton, 13
proximity detector (Jameco), 140

R

Radio Shack, 134
Radio Shack PCB kit, 119
radio with telephone hold (Jameco), 143
Radiokit, 133
reactance, 30
reactance transformer, 48
receivers, symbols for, 51
repulsion, law of, 14
resistance, dc, 30
resistance substitution kit (Jameco), 140
resistance-sensitive audio continuity tester, 179-181
resistors, 25, 38
    color code of, 26, 27, 28
    fixed, 45
    fusible, 46
    ohm as unit for, 26
    schematic symbol for, 45
    special-purpose, 46
    tolerance in, 27
    variable, 45
    watt sizes for, 26
reversing film, 118
robot blinker (EKI), 129
robot navigator (Jameco), 144
robot walker (Jameco), 144
robot wheel (Jameco), 143
robots (Jameco), 143
rosin, 72

S

safety, 9-12
    designing for, 11
schematic diagrams, 45
    audio amplifiers in, 51
    capacitors in, 46
    circuit components in, 52
    circuit connections in, 51
    current sources in, 50
    diodes in, 50
    inductors in, 47
    isolators in, 51
    linear IC amplifier circuit, 54
    miscellaneous symbols in, 49
    numeric designations on, 54
    plugs and jacks in, 50
    receivers in, 51
    resistors in, 45
    small-signal transistor circuit, 108

switches in, 49
symbol template for, 55
transfer to PCB of, 107-110
transformers in, 48
transistors in, 51
transmitters in, 51
understanding of, 53
vacuum tube in, 53
voltage sources in, 50
semiconductors, 19-21, 33-34
    diode, 33
    doped crystal, 21
    electrons in shells of, 20
shell, 17
shimmer lights (EKI), 129
Shockley, William, 34
signal generator (EKI), 128
silicon, 19, 34
siren oscillator (EKI), 129
siren/code oscillator (Jameco), 140
small-scale integration chips, 38
small-signal transistor circuit, 108
solder, 69
    melting temperature of, 70
    tin-to-lead ratios in, 70
solder lugs, 56, 57
solder sucker, 73
soldering, 56-85
    aids in, 65
    components, 124
    copper tips for, 64
    flux in, 71
    iron-plated tips for, 64
    printed wiring boards and, 79
    repairing damaged or missing PCB tracks, 82
    rosin, 72
    steps in, 68-72
    tinning in, 78
    tips for irons, 62
    tools for, 60
    wire and terminal preparation for, 80
    wraps and splices, 81
soldering irons, 60
solderless connectors, 56, 57
sound effects generator (Jameco), 140
sound experimenter (Jameco), 138
sound flasher (Heathkit/Zenith), 146
sound-operated robot (Jameco), 143
sound-operated switch (Jameco), 142
space war gun (EKI), 128
special-purpose resistors, 46
splices, 82
    soldering of, 81
stop-action timing tester (EKI), 132

strobe stick (EKI), 130
substrate wafer, 36
surface mount technology (Heathkit/Zenith), 149
switches, symbols for, 49

T

TEC-200 Image Film, 121-122
    etching with, 121
    pattern transfer with, 121
    photocopying the circuit with, 121
telephone hold button (EKI), 132
telephone hold with music (Jameco), 142
telephone horn button (Jameco), 142
telephone on-hold, 195-197
temperature-controlled soldering station, 60
templates, 55
terminal strips, 59
terminals, 57, 58
    soldering preparation for, 80
testing PCBs, 126
tinning, 78
tolerance, resistors, 27
tools, 6-9, 56-85
transformers
    air-core, 48
    iron-core, 48
    reactance, 48
    schematic symbol for, 48
transistor/diode checker (Jameco), 141
transistors, 34, 37
    symbols for, 51
transmitters, symbols for, 51
triodes, 32, 33
trivalent elements, 21
tunable electronic organ (Jameco), 140
TV jammer (EKI), 130
TV sound enhancer (Jameco), 142
two-IC radio (Jameco), 142

U

ultralarge scale integrated circuits, 38
ultrasonic transmitter and receiver, 185-191

V

vacuum tubes, 31
    symbols for, 53
variable power supply(Jameco), 140
variable resistors, 45
variable strobe light (EKI), 130
variable strobe light (Jameco), 139
very high speed integrated circuits, 38

very large scale integration chips, 38
visual telephone ringer, 197-199
voltage, 25
    drop in, 25
    source of, 25, 50

W
warbling siren (Jameco), 140
whistler lighting receiver, 169-173
whopper alarm (EKI), 130
whopper siren (Jameco), 140

wire cutters, 7, 60
wire wraps, 66, 81
wireless mike (Heathkit/Zenith), 146
wires
    color code for, 87
    copper, specifications for, 88
    designations for, 87
    sizes for, 87
    soldering preparation for, 80
wiring
    breadboarding as, 89-93

diagrams for, 44
point-to-point, 89
workshop
    hints for, 4-6
    safety in, 9-12
    tools and equipment for, 6-9
wrenches, 7

Z
zero-to-nine-volt power supply (EKI),
    129

Other Bestsellers From TAB

Other Bestsellers From TAB

☐ **HOW TO DESIGN AND BUILD ELECTRONIC INSTRUMENTATION**—2nd Edition—Joseph J. Carr

One of the most useful books ever published on the design and construction of electronic circuitry has now been completely revised to include a wealth of applications devices . . . including the latest in microcomputer-based instrumentation and actual computer programs that are sure to make designing circuits less complicated. 518 pp., 328 illus.

Paper $19.95   **Book No. 2660**

☐ **THE ENCYCLOPEDIA OF ELECTRONIC CIRCUITS**—Rudolf F. Graf

Here is every professional's dream treasury of analog and digital circuits—nearly 100 circuit categories . . . over 1,200 individual circuits designed for long-lasting applications potential. Adding even more to the value of this resource is the exhaustively thorough index which gives you instant access to exactly the circuits you need each and every time! 768 pp., 1,762 illus.

Paper $39.95   **Hard $60.00**
Book No. 1938

☐ **1001 THINGS TO DO WITH YOUR IBM PC®**—Mark R. Sawusch and Tan A. Summers

Here's an outstanding sourcebook of microcomputer applications and programs that span every use and interest from game playing and hobby use to scientific, educational, financial, mathematical, and technical applications. It provides a wealth of practical answers for anyone who's ever asked what can a personal computer do for me? Contains a gold mine of actual programs and printouts, and step-by-step instructions for using your micro. 256 pp., 47 illus.

Paper $10.95   **Hard $12.95**
Book No. 1826

☐ **IBM PC® GRAPHICS**—John Clark Craig and Jeff Bretz

Now, this practical and exceptionally complete guide provides the answers to questions and the programs you need to utilize your IBM PC's maximum potential. This is a collection of immediately useful programs covering a wide variety of subjects that are sure to captivate your interest . . . and expand your programming horizons. 272 pp., 138 illus., 8-page color section.

Paper$14.95   **Hard $16.95**
Book No. 1860

☐ **BUILD YOUR OWN LASER, PHASER, ION RAY GUN AND OTHER WORKING SPACE-AGE PROJECTS**—Robert E. Iannini

Here's the highly skilled do-it-yourself guidance that makes it possible for you to build such interesting and useful projects as a burning laser, a high power ruby/YAG, a high-frequency translator, a light beam communications system, a snooper phone listening device, and much more—24 exciting projects in all! 400 pp., 302 illus.

Paper $16.95   **Book No. 1604**

☐ **33 GAMES OF SKILL AND CHANCE FOR THE IBM PC®**—Robert J. Traister

Turn your IBM Personal Computer into a super computer arcade during its off-work hours! Here's a collection of challenging and entertaining games that can provide hours of enjoyment, *plus* help you extend your computing skills and build your programming knowledge. Includes brain teasers, games for amusement and relaxation, educational games, and games for every age level! 256 pp., 44 illus.

Paper$14.95   **Hard $18.95**
Book No. 1526

Send $1 for the new TAB Catalog describing over 1300 titles currently in print and receive a coupon worth $1 off on your next purchase from TAB.

▬▬▬▬▬▬▬▬▬▬▬▬▬▬▬▬▬▬▬▬▬▬▬▬▬▬▬▬▬▬

To purchase these or any other books from TAB, visit your local bookstore, return this coupon, or call toll-free 1-800-233-1128 (In PA and AK call 1-717-794-2191).

Product No.	Hard or Paper	Title	Quantity	Price

☐ Check or money order enclosed made payable to TAB BOOKS Inc.

Charge my   ☐ VISA   ☐ MasterCard   ☐ American Express

Acct. No. _____ Exp. _____

Signature _____

Please Print

Name _____

Company _____

Address _____

City _____

State _____ Zip _____

Subtotal	
Postage/Handling ($5.00 outside U.S.A. and Canada)	$2.50
In PA add 6% sales tax	
TOTAL	

Mail coupon to:

TAB BOOKS Inc.
Blue Ridge Summit
PA 17294-0840   BC